Careers in Food Science: From Undergraduate to Professional

Richard W. Hartel · Christina P. Klawitter
Editors

Careers in Food Science: From Undergraduate to Professional

Editors
Richard W. Hartel
University of Wisconsin
Madison, WI
USA
rwhartel@wisc.edu

Christina P. Klawitter
University of Wisconsin
Madison, WI
USA
cklawitter@cals.wisc.edu

ISBN: 978-0-387-77390-2 e-ISBN: 978-0-387-77391-9
DOI: 10.1007/978-0-387-77391-9

Library of Congress Control Number: 2008929639

Printed on acid-free paper

springer.com

Preface

Do you like food? Do you like science? How about applying science to food? Consider food science for a career. The range of career options with a degree in food science is so broad that you can do almost anything you'd like once you graduate.

The overall goal of this book is to help you explore the wide range of exciting and interesting careers available to those who hold food science degrees. To help with career management, the book has been broken into sections appropriate for different stages of development in a food science career. Part I provides a brief introduction to careers. Although the focus is on food science careers, the advice found in Part I applies, no matter what field you choose. In Part II, we've incorporated chapters about the undergraduate experience. From choosing a major, pursuing internships, utilizing university resources, and using your time as an undergraduate to develop a host of professionally relevant skills and experiences, almost every facet of the undergraduate experience is covered by people who have either just gone through that experience or whose job is to provide assistance during that experience. Guidelines for choosing a graduate school, finding a major professor who suits your needs, and getting a research career off the ground are covered in Part III. In Part IV, people with years of experience provide advice on how to find a job in the food industry and, once you have a job, how to ensure that you have a successful career. Part V provides extensive information about a host of potential career paths students with food science degrees can take after earning both undergraduate and graduate degrees. The chapters in Part V demonstrate the breadth of the food industry, what work in the industry is like on a daily basis, and provide insight into the requisite skills for success, and what an undergraduate can do to prepare for a career. These chapters were written by people with successful careers in that job function. Finally, Part VI provides insight into academic careers, from finding a position to getting tenure.

The authors of these chapters were invited to contribute to this book based on their experience with the career decision-making process and their expertise within the food industry. All were students at one time who faced career decisions, much like the decisions you may be facing in the near future. We hope their descriptions and advice will be useful tools for you as you hone in on the best way to focus your professional interests and skills in the food industry.

Madison, Wisconsin

Richard W. Hartel
Christina P. Klawitter

Contents

Contributors

Leann Barden
University of Wisconsin, Madison, WI, USA

Katie Becker
Attorney at Law, Chicago, IL, USA

Alison Bodor
National Confectioners Association, Vienna, VA, USA

Richard Boehme
Kerry Ingredients, Beloit, WI, USA

Angela Byars-Winston
UW Center for Women's Health Research, Madison, WI, USA

Rose Defiel
Clasen's Quality Coatings Middleton, WI, USA

Melody Fanslau
Fair Oaks Farms, LLC, Pleasant Prairie, WI, USA

Laura Folts
University of Minnesota, St. Paul, MN, USA

Richard W. Hartel
University of Wisconsin, Madison, WI, USA

Susan Hough
The Masterson Co., Milwaukee, WI, USA

Becky Kuehn
University of Wisconsin Madison, WI, USA

Christina Klawitter
University of Wisconsin, Madison, WI, USA

Kalmia E. Kniel
University of Delaware, Newark, DE, USA

Dennis Lonergan
The Sholl Group II, Inc., Eden Prairie, MN, USA

Silvana Martini
Utah State University, Logan UT, USA

Moira McGrath
OPUS International, Inc., Deerfield Beach, FL, USA

Brian McKim
Kraft Foods Inc., Glenview, IL, USA

Cassandra Miller
SUSTAIN, Washington, DC, USA

Jennifer Neef
University of Illinois, Urbana, IL, USA

Suzanne Nielsen
Purdue University, West Lafayette, IN, USA

Christine Nowakowski
General Mills, Minneapolis, MN, USA

Kelsey Ryan
Purdue University, West Lafayette, IN, USA

Leslie Selcke
University of Illinois, Champaign, IL, USA

Michelle Tittl
Target Corporation, Minneapolis, MN, USA

Peter Weber
Potter's Crackers, Madison, WI, USA

Janelle Young
Lactalis, Belmont, WI, USA

Dennis Zak
TMResource LLC, Doylestown, PA, USA

Tanya Zimmerman
University of Wisconsin, Madison, WI, USA

Part I
Introduction

Chapter 1
Introduction: Career Preparation for the Food Industry

Richard W. Hartel

There are over 6.6 billion people on the planet, and every one of us has to eat. Although people in some parts of the world (from the poor in Africa to the homeless in New York City) suffer from malnourishment, the majority of people in the world have more than enough to eat. Making sure that all these people eat every day makes the food industry one of the largest in the world.

In addition, it's an exciting time to be involved with food and its production. The popularity of the Food Network, and the broad range of food shows on the television, clearly indicates the importance we put on food. Moreover, recent advances in nutrition, biochemistry, medicine, and many more fields, have spurred an interest in food as a vehicle for nutrition and health. The adage that "you are what you eat" has never been truer today than at any time in the past. You can still enjoy a Twinkie if you so desire (remember, you are what you eat), but the majority of the food industry is heading toward providing optimal nutrition to enhance health and wellness. And, doing it in an environmentally friendly way.

Almost no matter what you want to do, the food industry offers you the opportunity because of its size and diversity. If you're a Food Science major graduate, or thinking of majoring in Food Science, the food industry is your oyster, so to speak. The future is bright and you have the chance to be a part of one of the largest and most important industries around the world. You can satisfy your desires to do good in the world while still have an enjoyable and profitable career.

R.W. Hartel
University of Wisconsin, Madison, WI, USA

R.W. Hartel, C.P. Klawitter (eds.), *Careers in Food Science*,
DOI: 10.1007/978-0-387-77391-9_1,

The Vast Food Industry

A walk through the grocery store aisles shows the diversity of food products available these days. For nearly every food, from peanut butter to yogurt, there are literally dozens of options to choose from. Where do all these foods come from? The "food industry."

But what exactly do we mean by "the food industry?" Most of us think of the larger food manufacturers, from Kraft to Hershey, but there's so much more. There are suppliers of ingredients, equipment, and even personnel. There are small companies that make foods for niche markets. There are the government regulatory agencies that ensure the safety of the food supply. And, there are training, analysis, and research facilities, from universities to private labs to government labs, which support food manufacturers and suppliers of all sizes. It is this wide variety of important components that makes the food industry so large and diverse, resulting in the broad range of job opportunities. Let's look at some of the opportunities in a little more detail.

Food manufacturers are those companies that turn raw materials (or intermediate products) into finished products – the ones we see available on the grocery store shelves. Food manufacturers vary in size from small startup companies (think of Ben and Jerry making ice cream in their garage) to the largest multinational conglomerates (Nestle is the largest food company and one of the largest ice cream manufacturers), with every size of company in-between. Of course, these food manufacturers hire food scientists for a wide range of jobs. There are production supervisors (see Chapter 19), who oversee the manufacturing lines, and plant managers, who oversee the entire operations within a manufacturing facility. Quality control/assurance people (Chapter 18) ensure the products are safe to eat and meet the company's quality standards. Product developers (Chapter 20), often in conjunction with research chefs (Chapter 26), are the people who create and develop the new products we see on the grocery store shelves. The job of research and development (R&D) personnel (Chapter 22) is to better understand the physical, chemical, and microbial attributes of foods, and to develop new and better manufacturing technologies. Although marketing is not typically a job that food scientists fill immediately upon graduation, some food scientists gravitate toward the business side of food manufacturing and transfer into the marketing division after a few years. There are even food scientists who start their careers as purchasing agents within a manufacturing company; they are responsible for procuring the necessary ingredients from supplier companies.

Suppliers are companies that provide ingredients or components to the food manufacturers. One of the largest supplier industries is the flavor industry. A food scientist at a flavor company works directly with the food manufacturers to find exactly the right flavors that suit a new product being developed. For example, when Derek, a product developer at an ice cream company, decides to make a new flavor of ice cream, he approaches, Chris, the technical sales representative (see Chapter 21) at a flavor company with an idea. Chris takes the idea back to the flavor labs, where flavor scientists develop prototypes for evaluation by Derek. Chris and Derek may go back and forth several times until Derek feels he has the right flavor to suit his new ice cream, at which point they go into plant trials and eventually, if everything is successful, into full-scale production. If the new flavored ice cream makes it through all the hurdles to get to market, eventually consumers like you and I can buy it at the grocery store. And both Chris and Derek get a huge feeling of satisfaction when they see their product on the grocery store shelves.

Ingredient or food component supplier companies vary widely. Ingredient suppliers provide food components, such as flavors, colors, acids, specialty chemicals, and other ingredients. Some supplier companies produce intermediate ingredients for food manufacturers to use. For example, there are large cookie and biscuit manufacturing companies who don't make any chocolate themselves. They simply buy chocolate as an ingredient from chocolate supplier companies. In the same way, as Derek and Chris worked together to come up with an appropriate flavor, food scientists John and Michelle worked together to ensure that Michelle's chocolate coating would work well on John's new cookie product.

Another example of an intermediate ingredient supplier is the company that makes breading for frozen breaded onion rings made at a local manufacturing plant. Food scientist Jeanne from the frozen onion ring manufacturer would work with Lori, a product developer at the breading ingredient supplier, to make sure the breading had exactly the right qualities at a low enough cost to fit within the profit window.

Other suppliers of food manufacturers include companies that provide packaging materials and manufacturing equipment. Some food scientists may gain a specialization in packaging, bringing a deeper knowledge of how package and food interact to enhance safety, convenience, and marketing of products.

Other companies or organizations that either support or interact with food manufacturers include *special interest groups* (Chapter 24), *regulatory agencies* (Chapter 23), and *service companies* (i.e., companies that provide chemical and/or microbial analyses, external sensory labs, etc.). One could also include *placement agencies*, or companies that specialize in matching food scientists looking for employment with companies looking to hire new employees, in the list of companies that support the food industry.

According to the Food and Agriculture Organization (FAO) of the United Nations (UN), our food supply is easily sufficient to feed the entire population of the world. Advances in crop production and yields have continued to improve food production at a pace fast enough to keep up with the growing population. However, at least 800,000 people around the world, mostly from developing countries but also some here in the US, are undernourished. If you're interested in helping to ensure that everyone in the world gets enough nourishment every day, there are also opportunities to use a degree in Food Science to help the public good (Chapter 25). Consider working in the Peace Corp or at various nonprofit institutions whose aim is to provide healthy and nutritious foods where needed around the world.

Perhaps you're interested in starting your own business (Chapter 27)? The food industry is an interesting economic model, with large companies seeking growth through purchase of smaller companies. That provides an opportunity for new startup companies to fill a niche in the market. A degree in Food Science, with a business slant, provides an excellent background for becoming a food entrepreneur.

Again, there is a wide diversity of opportunities available for employment within the food industry. Almost no matter what your inclination for a career, you can find an opportunity for it in the food industry.

Prepare for a Career in the Food Industry: A Learner's Guide

This book is chock full of tidbits on how to prepare for almost any job related to foods. However, one of the most important things to get out of this book is the message that it's all up to you. You can make what you want of life, no matter where you are right now.

If you're a new student, just considering a major in Food Science, you can find some great advice in Part 2 to help you make the most of your college career and Part 5 detailing numerous job opportunities in the food industry. If you're just completing a BS degree in Food Science and considering graduate school, Part 3 provides guidance for finding an appropriate graduate program, making the transition to graduate school and making the most of the opportunity.

For many, finding and starting your first job in the industry can be quite daunting. What is expected of me? How should I behave? What types of things should I watch out for? How do I make the most of the opportunities presented? Part 4 is intended to provide guidance about how to make the most of your opportunities. And finally, for those who have made it through the seemingly infinite hoops and hurdles to face an academic job, Part 6 provides insight into one of the most difficult, yet one of the most rewarding, opportunities – the academic job.

Almost no matter what your inclination within the food industry, you can find insight and advice from past experience within this book.

Why should you consider the food industry for your career? Again, the wide diversity of opportunities in the food industry allows you to find your place almost no matter what your career aspirations. Do you like science with a meaningful application? Are you interested in psychology – why do people eat what they eat and taste what they taste? Are you creative and want to produce new and exciting foods? Are you concerned about people's health and want to do something about it? Do you make a killer cookie – maybe you can turn that skill into a successful business? The opportunities are virtually endless with a degree in Food Science.

Chapter 2
The Equilibrium of Life–Career Planning

Angela Byars-Winston

Introduction

Over a 100 years ago, fixed wing aircraft flight was made possible by Orville and Wilbur Wright's breakthrough invention of a three-axis control that allowed effective steering and maintenance of airplanes' equilibrium. The significance of the Wright brothers' control invention is that it facilitated recovery of lateral balance, or equilibrium, when the wind tilts the aircraft to one side or another. A structure is in *equilibrium* when all forces acting on it are balanced. Thus, equilibrium can be thought of as recovery of force balance.

As a college student, you are no doubt balancing many forces: academic, social, and personal. These may include managing increased class workloads, relations with faculty and staff, and meeting new people. In learning to balance these forces and finding your equilibrium, the insights of a great life strategist named Dr. Seuss is useful. In his book, *Oh, the Places You'll Go*, he eloquently described life as the "great balancing act" full of adventure and challenge. How appropriate that this was one of the last books he would author. It was published in 1990 and he passed away in 1991. Dr. Seuss penned this work from a deep place of wisdom, looking back over years of living and the lessons that his life journey had provided.

Life indeed is a balancing act in which you find equilibrium between your life and career. This chapter will focus on strategies for balanced life–career planning. In this chapter, I will (1) present a rationale for life-career planning, (2) discuss career beliefs that influence the career choice process, and

A. Byars-Winston
Institute for Clinical and Translational Research, School of Medicine and Public Health, UW Center for Women's Health Research, 700 Regent Street, Suite #301, Madison 53715-2634, WI, USA

R.W. Hartel, C.P. Klawitter (eds.), *Careers in Food Science*,
DOI: 10.1007/978-0-387-77391-9_2, © Springer Science+Business Media, LLC 2008

(3) discuss specific strategies for engaging in a reflective approach to choosing a career. The chapter concludes with resources that can further assist you in your planning process.

Why a Life–Career Planning Focus?

Howard Gardner, a professor of cognition at Harvard University, advanced the notion of *multiple intelligences* proposing diverse ways in which individual competence can be demonstrated. For instance, although the formal educational system of the United States largely values and rewards competence in what Gardner describes as logical-mathematical intelligence and linguistic intelligence, he asserts that human capacities can range from intelligence involved in interpersonal functioning to musical intelligence. One exciting and liberating aspect of Gardner's theory is the implication that if all humans have multiple capacities then we also have multiple potentialities.

The relevance of multiple intelligence theory to life–career planning is that there are many pursuits to be considered in which you may realize your particular competencies. This is important to keep in mind given that the labor market is constantly changing. To be sure, the number of jobs that people will hold over time is going up as the duration of their employment is going down. A survey of workers found that 26% of workers were in their current job for 12 months or less; only 15% were in the same job for 3–4 years (McGinn and McCormick, 1999).

There are just as many paths to becoming a food scientist, or an agronomist, dietitian, horticulturalist, botanist, etc. as there are paths to practicing in those fields. Do not focus exclusively on securing a particular job, but rather consider the *types* of work you might enjoy. Whereas a job is your current employment position, your career is the total sequence of those jobs over your total work history. You are well served to consider your career as a dynamic and evolving process that will unfold across your lifespan.

I could not discuss life–career planning without addressing the role of doubt in this process. Indeed it is one of the most common forces that disturbs the life equilibrium of a college student, leaving one feeling out-of-balance and just plain stressed. Have you felt embarrassed to admit that you had doubt about your career choice? Have you felt absolutely sure about your career goal at one point only to experience something new and second guess that goal? Let me encourage you to see this uncertainty as normal. In fact, the uncertainty you feel reflects most people's reality. Did you know that, on average, about 50%–54% of first year college students indicate that they expect to change their college major or career plan at least once during their college experience (Hurtado et al., 2007)? In national surveys of Americans over age 18, most indicate that they are unsure of how to make informed career choices (NCDA Gallup poll, 1988). Although there is a segment of the college population that is sure about their college major and eventual career path and actually pursue those, for many the only certainty about choosing a career is the uncertainty that comes with the choice process.

In a larger sense, the certainty of any given choice is judged only in retrospection by its outcome. We truly do not know whether a choice we made was correct until we experience the consequences. It is really in the searching and the uncertainty that we learn. Thus, I encourage you to pursue your food science major not based on certainty of a particular career choice but based on certainty of personal development.

For some, the uncertainty and ambiguity of career choices may feel unsettling. As a professor, I advised scores of students regarding their program of study and career choices. I cannot tell you how often they were in a state of anxiety at best or a state of panic at worst over the uncertainty of choosing both. My counsel was always the same: do not rush into a decision.

It is important to engage in reflection on what life you want to live and then explore what types of work would be compatible. Nonreflective people ask, "What could be worse than unanswered questions?" Reflective people reply, "Unquestioned answers." Take time to reflect on who you are and who you want to become.

Before I discuss specific life–career planning strategies, I would like to address some dynamics particularly related to choosing a career that can affect the planning process. The next section of this chapter briefly considers the influence of career beliefs on your academic and career-related choices.

Career Myths and Career Realities

Chances are, once you told people that you were going to college, the question they likely asked was, "What are you going to major in?" After that, the next likely question they asked was, "What are you going to do after college?" If you ask a group of seniors, "Are you majoring in what you thought you would major in as a freshmen?" most will say no. Some will tell you they never even heard of their major prior to attending college. If you are like many college students, of you have been in school continuously since you were 5 or 6 years old. With much of your life dedicated to being a student, you may not have a broad sense of the 20,000+ jobs in the United States identified by the Department of Labor. Knowledge about careers can be based on myths and stereotypes as well as factual information.

Let us consider some of the myths about choosing and starting a career. Here are some examples, a few of which were identified by Crosby (2005).

Myth 1: "There is one perfect job for me."
Myth 2: "My college major will determine what career I have."
Myth 3: "I have to use all of my abilities in my job."
Myth 4: "Once I choose a career, I cannot change it."

Sound familiar? These beliefs about careers and how to get started in a career can stifle your search process and discourage you from exploring who you are and what work might fit you. If you feel like you have to major in food science because you want to work in the food industry after graduation, here are some interesting facts to correct the aforementioned myths that may put the process of choosing a career into perspective.

Reality 1: There are many jobs, and possibly careers, you might enjoy. Since your interests and preferences might change over time, it is wise to keep your options open.
Reality 2: A national longitudinal survey of over 9,000 people who completed bachelor's degrees in 1992–1993 found that 4 years after obtaining their degree, only 55% were in jobs related to their major (Horn and Zahn, 2001). Moreover, only 58% of this group reported that their undergraduate major field remained important to their lives 10 years after graduation (Bradburn et al., 2006).
Reality 3: Because you have multiple abilities and talents, no single job will likely allow you to incorporate all of them. However, each job you acquire will allow you to hone your skills for future employment.
Reality 4: College graduates often change fields once they enter the job market or later in their career (Henke and Zahn, 2001).

Despite the numerous misconceptions about careers that exist, the career reality is that many people will not know exactly what they want to do until they try it out in college or after college. Interestingly, 69% of those surveyed in a national Gallup Poll indicated if they could start over in choosing their career, they would get more information about career options than they did the first time (Kaplan, 2000). Whether you are certain of pursuing a food science career or you feel like a ship without a sail, there are practical steps you can take to advance your life–career plan. Whereas other chapters in this edited book will give you specific information on ways to conduct a successful job search in the food science industry and how to use career service centers, the next section of this chapter outlines general life–career planning strategies.

Practical Tips for the Outstarting of Your Journey: Back to Dr. Seuss

The book *Oh, The Places You'll Go* is a call for each person to find her or his way in life, persisting through challenges to realize one's success. This focus is consistent with life–career planning in that your career choice process will develop over your lifetime, continually influenced by the various life experiences you have and how you grow from each of them.

There are many models and approaches to career planning that can aid you in your choice process. One commonly used model includes reflection on three questions: Who Am I? Where Am I Going? How Do I Get There? I find these questions to be intuitive and easy to follow. In Table 2.1, I have extended the model to the two dimensions of life and career to encourage you to reflect on these three questions in both domains. Presented in the following text are seven strategies that are in no sequential order but are meant to be guideposts for you to consider as you plan your life and career.

Table 2.1 Applications of life–career planning approach to three-step model

	Life	Career
Who am I?	• What are your values? What do you think is fun and interesting? • What is your sense of purpose? • Consider how you spend your free time; what does this suggest about your abilities, interests, values? • What do you enjoy learning about? • What are you naturally suited for?	• In what areas are your skills the strongest: working with people, data and ideas, or things? • What kind of people might you like to be with 10 hours each day? • Does your work need to be meaningful? • What kinds of careers do you regularly think about? That is, what careers do you often daydream about? • What kinds of rewards do you need in work to remain satisfied?
Where am I going?	• How long am I willing to take to develop my career and get the job position I want? • Do I have a time limit for entry into the career I want?	• What occupational information do I need? • What are the various paths into my career area? • Are there other ways into the job position?
How do I get there?	• Make a specific life and career goal statement	• Identify strategies, methods to obtain successful job entry

Table 2.2 Examples of life–career planning strategies

	Life–personal identity	Career identity
1. Anticipate challenges and prepare "Bounce Backs"	Construct a list of "Plan Bs"	Consider and explore alternate career paths and alternate job positions within your field
2. Develop good social networks	Maintain friendships, precollege mentors (e.g., high school teachers)	Join as student member of professional associations; join mentor networks offered by associations
3. Develop your cross-cultural (in vivo) experiences	Understand your personal culture and worldview; expand your circle of friends and experiences across diverse cultures	Identify cross-cultural skills needed in and relevant to your career field

Table 2.2 provides examples of life–career planning considerations for three of the strategies.

1. Know Yourself. If there are two people on a journey and one has a map and the other does not, who is more likely to reach her or his destination? The one with the map. In order to get directions to a particular location through a global positioning system (GPS) or Mapquest.com, one has to know both her or his origin and destination. When I was a counselor at a university counseling center, many clients came to the center for career help. They wanted to "figure out" what they were going to do after graduation in a couple of counseling sessions or after one career assessment. I often had to convince them that understanding who they were as a person would help them better "figure out" a career for themselves.

Socrates stated that the unexamined life is not worth living. When you examine your life, who you are and how you got there, you are better able to set clear, informed goals and better evaluate when your career path is working for you or not. Examining your values is a good place to start understanding who you are—what is important to you and why? What motivates you? What do you expect will happen if you enter a food science career? Edward Colozzi (2003), a psychologist in private practice in the northeastern United States, has found that his clients who spend in-depth time on clarifying their life values make better career choices.

In examining who you are, also consider who you want to become. Specifically, it is important to think about what kind of life you want to have and the multiple life roles you might assume in the future (e.g., partner, parent, entrepreneur). Consider how a food science career will be compatible with those multiple roles. A practical strategy for self-awareness that

I often use in career planning is for students to conduct a SWOT analysis of their personal career (see Table 2.3). SWOT refers to analysis of one's Strengths, Weaknesses, Opportunities, and Threats. This will facilitate concrete thinking about who you are and aspects of yourself that bear on your career planning.

A word of caution: take care not to get stuck in introspection. Avoid the pitfall of paralysis-of-analysis. Insight must be accompanied by action and often times that insight comes by actually trying out and testing reality.

2. Develop Transferable Skills. A college degree, undergraduate or graduate, does not guarantee success in the workforce. It has been said that a college degree and a dollar will get you four quarters: you may know intellectually how to divide the dollar into various quantities, but not necessarily how to make it work for you. The U.S. Department of Labor Secretary's Commission on Achieving Necessary Skills (known as SCANS–http://www.wdr.doleta.gov/SCANS/) identified and described job skills necessary for success in the world of work. These skills include good communication skills (written and oral), ability to work well in groups, leadership, and ability to work with cultural diversity. The skills are often referred to as "soft skills" in that they are skills that are transferable to and useful in any field of work. They help to complement the "hard skills," or those discipline-specific skills, which are typically acquired through formal education and training. Transferable skills are the sine qua non factor of successful employment.

It is important to note that most of these skills will be learned and honed *outside* of the college classroom. In Bill Coplin's (2003) book *Ten Things Employers Want You to Learn in College*, he describes specific ways for college students to prepare for a career by developing 10 skills that he calls Know-How Skills (KHS), which mirror the SCANS skills mentioned earlier. One KHS is "Establishing a Work Ethic." Under that skill, Coplin identifies four skill sets to advance your proficiency in this area: (1) Self-initiative (a.k.a. "Kick Yourself in the Butt" skill), (2) honesty, (3) time management, and (4) money management. What is particularly useful about his book is that it includes worksheets for KHS self-assessments, calendars for developing KHS, and practical ideas for building KHS outside of the classroom. The Web site address for more information on the book and downloadable KHS worksheets is listed in Table 2.4.

3. Develop Good Social Networks. Many people find their current job based on who they know (not just through job ads). I am sure you can think of a person you know who got a job because they knew someone personally in the organization. Or perhaps someone you know was selected for an internship because her or his professor knew the person doing the hiring.

Table 2.3 SWOT personal career analysis and examples of SWOT questions (Adapted by Angela Byars-Winston, 2005)

	Your STRENGTHS: Internal positive aspects that are under your control and upon which you may capitalize in planning	Your WEAKNESSES: Internal negative aspects that are under your control and that you may plan to improve
INTERNAL	• What do you do well? • What need do you expect to fill within your work setting? • What knowledge or expertise will you bring to your desired work position? • What is your greatest asset?	• What are your professional weaknesses? • How do they affect your job performance? (These might include weakness in specific skill areas like leadership or interpersonal skills.) • Think about a career-related challenge you have experienced and consider whether some aspect of your personal or professional life could be a contributing factor to that experience(s)
	OPPORTUNITIES: Positive external conditions that you do not control but of which you can plan to take advantage	THREATS: Negative external conditions that you do not control but the effect of which you may be able to lessen
EXTERNAL	• Where are the promising prospects facing you? • What is the " state of the art" in your particular area of interest? • What are you doing to enhance your exposure to and development in this area? • What formal training and education can you or do you need to add to your background that might better position you appropriately for more opportunities?	• What obstacles do you face to realizing your career goals? • How can you maintain your interest in the face of delayed progress? • Could your area of interest be fading in comparison with more emergent fields? • Are there aspects of your desired work setting that will lead to conflict or particular challenges for you? • How might you work with or change those aspects to perhaps diffuse these threats?

Adapted from: Quintessential Careers, http://www.quintcareers.com/SWOT_Analysis.html

Personally, I was introduced to psychology research and academia as a career path as an undergraduate after being hired to enter data on a professor's research project. That professor noticed my research interest, mentored me into a graduate program, and remains a close friend today after nearly 20

Table 2.4 Useful career resources

- University/College Career Services – see chapters in Section 2 of this edited book: (career advising and career assessment (individual and group services, workshops, etc.), solid career resources, useful publications like "What Can I Do With a Major In...?" lists)
- University/College Counseling Center may provide individual support to explore life–career concerns and in-depth personal support when needed
- Family members (useful resources to inform your SWOT Personal Career Analysis)
- Faculty/staff
- Bill Coplin's (2003) book, *"10 Things Employers Want You to Learn in College: The Know-How You Need to Succeed."* Published by TenSpeed Press.
- Web site to order book and download worksheets: http://www.tenspeed.com/store/index.php?main_page=pubs_product_book_jph1_info&products_id=1628.
- Gregg Levoy's Callings Web site: http://www.gregglevoy.com/ (useful references like movie titles to prompt reflection on your values)
- U.S. Department of Labor and Bureau of Labor Statistics websites:
  - Free online career self-assessments: http://www.online.onetcenter.org/
  - Matching skills to occupational fields: http://www.bls.gov/opub/ooq/2004/fall/art01.pdf
  - "Career Myths and How to Debunk Them:" http://www.bls.gov/opub/ooq/2005/fall/ art01.htm
  - Informational Interviewing: http://www.bls.gov/opub/ooq/2002/summer/art03.pdf
  - The National Career Development Guidelines (indicators of career competencies): http://www.acrna.net/files/public/ncdg.doc

years. I would not have considered an academic career without her encouragement. Clearly, you should develop good relationships with faculty and staff—they will give you academic support in completing your degree, but equally importantly, will help with practical support such as providing letters of recommendation.

Family members and friends may become especially important sources of support and potential resources as you advance your career planning. A recent study my colleagues and I published found that for undergraduate life science majors, their families were the biggest source of perceived support as they prepared for medical school (Klink et al., 2008). Moreover, family, friends, faculty and staff, and other trusted people in your life may provide good input for you on possible careers to consider. You may consider having friends or family members complete a SWOT analysis about you to gain some insight into your career planning. I also encourage you to consult people outside of your close circle (e.g., people in professional associations) for career input as those close to you may pigeonhole you into careers consistent with what they already know you like.

4. Dream Big/Take Risks. Given the prevalence of career myths, many people prematurely discard possible career options for themselves. This may be due to stereotypes about the types of people who work in a given career field or the tasks related to specific jobs within that field. Keep your career options broad as one never knows what opportunities might present themselves. The world is replete with stories of people who by serendipity happened upon a career that fit them or an incredible job opportunity. One survey of older adults found that 63% of men and 57% of women reported their careers were influenced by serendipitous events (Betsworth and Hansen, 1996). Unplanned circumstances can open up career possibilities previously unconsidered. At this point, you may be thinking that I am encouraging you to wander aimlessly and footloose in your career planning. After all, you have already focused on food science as a major and possibly a career. But that is not the case at all.

Since change is constant, in work and life, individuals are wise to set goals and make decisions in light of these dynamic contexts. Gelatt (1989) encourages people to incorporate the concept of "positive uncertainty" into their decisions, remaining focused on the process and journey of one's goals as much as on the outcome of that goal. He emphasizes that serendipity is "when you discover something good while seeking something else." To be sure, serendipity is not simply "a happy accident; it requires that you seek something and be receptive to something else." It requires that you take risks by following your intuition, being receptive to alternatives, and remaining open to learning. Life–career planning is a balance between setting goals and making plans that are informed by your self-knowledge and dreams mixed with unexpected circumstances that blend together and set you on a career journey.

5. Anticipate Challenges and Prepare "Bounce Backs." After the initial excitement of starting a new journey, the character in *Oh, The Places You'll Go* soon experiences various life events, or forces, which cause him to lose his equilibrium. These "bang ups" and "hang ups" leave the character feeling lost "in a lurch." He persisted on his journey, moving mountains and eventually found places "where Boom Bands are Playing.". Similarly, each life experience will help you further develop a personal identity in which a career identity can be folded.

Many students I advised and counseled were often demoralized by academic and personal challenges they encountered while pursuing college degrees. Oftentimes, they were most dejected by the fact that they had not expected a particular challenge to happen to them; in short, they did not see it coming. I encourage students to anticipate challenges and create plans of actions to manage them. I call these "bounce backs" or recovery plans. It is hard to rally self-assurance in the middle of unexpected crises; you are better off if you know what your coping resources are ahead of time. Learn how to manage and cope with change and challenges. Good coping strategies and *confidence* in your ability to engage those coping strategies (i.e., coping efficacy) are key to your life–career success.

One way to build your coping skills and efficacy is to write a week's journal of "Plan Bs." For one week, write down each day, on a sheet of paper, five decisions you made that day in Column A. In Column B, write down what you would have done had your decision in Column A *not* worked out. Deliberately reflecting on alternatives will (1) increase your awareness of multiple ways to achieve your goals and (2) establish a habit of making decisions from multiple options.

6. Develop Your Cross-Cultural Competence. The increasingly global economy requires that workers be culturally aware of who they are and the world around them. One of the important transferable SCANS skills employers look for is comfort with and ability to work across cultural diversity. Although some students may see this skill set as irrelevant to their job prospects, consider the following news story. In 2004, major corporations including Proctor & Gamble, Cargill, and Kimberly-Clark significantly scaled down recruitment at the University of Wisconsin-Madison citing that their graduates were "culturally incompetent," unprepared to work with people different from themselves (see editorial by Koehler and Miranda, 2004). To be sure, career success necessitates that employees be able to work well with others, not solely those with whom they are familiar.

An important dimension of the SCANS skill for cultural diversity is that competence in this area be *demonstrated.* I was the director of doctoral admissions in a graduate program for a few years. The program is nationally recognized for its multicultural training. Annually, we read over 80 applications, from which approximately 10–15 were selected for admission. There were many applicants who said they valued multiculturalism and diversity. But which applicant do you think stood out more: the one who said they valued multiculturalism or the one who practiced it and could demonstrate that value? We know that attitudes do not correlate perfectly with behavior, so it makes sense to choose the one who has had actual experience.

College students must do more than espouse cross-cultural values; they must have direct (in vivo) experiences. This may include participating in study abroad opportunities, learning another language (at least at the conversational level), or taking an informative cultural studies course. Deliberately expand your circle of friends to include more diverse people (e.g., language, geography) and diverse experiences (e.g., food, music, traditions). Coplin's (2003) book includes other suggestions for expanding cultural competence such as broadening your knowledge of various genres of music. It is also important to understand yourself as a cultural being regarding how your worldview has been informed by your family heritage, geography, gender, social class, ability status, race or ethnicity, and other cultural variables.

7. Be Informed. As mentioned previously regarding the influence of career myths, many people make career decisions based on what they *think* and not what they actually *know*. I encourage you to get good career information about the food science industry, employment projections, and job opportunities. In Table 2.4, I have listed numerous resources and Web sites that will prove useful to your career planning. Consider visiting the Bureau of Labor Statistics website in particular. They maintain current data on employment trends such as the Occupational Employment Statistics survey (http://www.bls.gov/oes) regarding employment at the national and local levels and the National Compensation Survey (http://www.bls.gov/ncs) that provides detailed wage information. The combination of self-knowledge and occupational knowledge will get you far.

Conclusion

Research with college and adult populations demonstrates that cheerfulness, or a positive disposition, is a strong positive influence on life satisfaction whereas feelings of depression are negative influences (cf. Shimmack et al., 2004). This research suggests that a lack of meaning is more detrimental to life satisfaction than stress and worries and, conversely, a cheerful temperament is more important than having a full social calendar. Take the time to know yourself and what is important to you. And finally, keep your life–career lens in the proper focus. Remain optimistic and encouraged that you will find your way in life and work. You are more likely to experience personal satisfaction if you have a positive outlook on your career choice process.

In the words of Dr. Seuss (1990), "Today is your day! You're off to great places! You're off and away! . . .And will you succeed? Yes! You will, indeed ! (98 and ¾ percent guaranteed)."

References

Betsworth, D., and Hansen, J. (1996). The categorization of serendipitous career development events. *Journal of Career Assessment, 4*, 91–98.

Bradburn, E.M., Nevill, S., and Cataldi, E.F. (2006). *Where Are They Now? A Description of 1992–1993 Bachelor's Degree Recipients 10 Years Later* (NCES 2007–159). U.S. Department of Education. Washington, DC: National Center for Education Statistics. Retrieved December 7, 2007: http://www.nces.ed.gov/pubs2007/2007159a.pdf.

Colozzi, E. (2003). Depth-oriented values extraction. *Career Development Quarterly*, 52, 180–189.

Coplin, W. (2003). *10 Things Employers Want You to Learn in College*. Jen Speed Press: Berkeley, CA.

Crosby, O. (2005). Career myths and how to debunk them. *Occupational Outlook Quarterly*, fall issue, *49*(3). Retrieved online april 21, 2007 at: http://www.bls.gov/opub/ooq/2005/fall/art01.pdf.

Gelatt, H. B. (1989). Positive uncertainty: A new decision-making framework for counseling. *Journal of Counseling Psychology, 36*, 252–256.

Henke, R. & Zahn, L. (2001). *Attrition of New Teachers Among Recent College Graduates: Comparing Occupational Stability Among 1992–93 Graduates Who Taught and Those Who Worked in Other Occupations*, NCES 2001–189. U.S. Department of Education National Center for Education Statistics: Washington, DC.

Horn, L.J., and Zahn, L. (2001). *From Bachelor's Degree to Work: Major Field of Study and Employment Outcomes of 1992–1993 Bachelor's Degree Recipients Who Did Not Enroll in Graduate Education by 1997* (NCES 2001–165). U.S. Department of Education. Washington, DC: National Center for Education Statistics.

Hurtado, S., Sax, L., Saenz, V. et al. (2007). Findings from the 2005 administration of Your First College Year. Los Angeles: Higher Education Research Institute.

Kaplan, R. (2000). *Playing with numbers in a Gallup Poll: 1999 National Survey of Working in America*. Presented at NCDA Conference, Pittsburgh, PA.

Klink, J., Byars-Winston, A., and Bakken, L. (2008). Coping efficacy and perceived family support: Potential factors for reducing stress in premedical students. *Medical Education, 42*, 572–579.

Koehler, T., and Miranda, C. (April 11, 2004). Are today's UW-Madison graduates cultural klutzes? No incentive encourages awareness. *Wisconsin State Journal. 102*, pp. B1.

McGinn, D., and McCormick, J. (February 1, 1999). Your next job. *Newsweek, 133* 42–51.

Seuss, D. (1990). *On, The Places You"ll go*. Random House, NY.

Shimmack, U., Oishi, S., Furr, R., and Funder, D. (2004). Personality and life satisfaction: A facet-level analysis. *Personality and Social Psychology Bulletin, 30*, 1062–1075.

Chapter 3
Overview of Careers in the Food Science Field

Moira McGrath

A Food Science degree is a door opener to many areas of specialty in the food industry. If you have chosen food science as a field of study, then you have a scientific mind and an interest to learn new things. But what types of job positions would be available, and what is the job market? What's new on the horizon?

Those who complete a food science degree, BS, MS, or PhD, have several different paths they can take. Research and development (R&D) and quality assurance (QA) are the two largest fields; however, each of these fields covers many different subfields. Let's say there is a salty snack manufacturer in the town in which you want to live, and you have always wanted to work there. What kind of jobs would be available to you if you have a food science degree?

If you are innovative by nature, and have a creative mind, you might be interested in developing new products. Instead of a cheese flavored salty snack, you might have an idea for a hot and spicy snack. In the *research and development* department, you would work with a team of other food scientists, each of you having a different role. All of the team members need to know about every aspect of product development, but each team member has an area of expertise. One person manages the innovation side of the project. He/she is the "idea person" with the original formula for the hot and spicy snack. Another handles the *microbiology* of the product. (Do bacteria grow on the snack product if the bag of chips is left on the grocer's shelves for more than three weeks?) Someone else handles the *food safety* of the product. (What ingredients should be used to assure that the product is safe to eat? Does mixing X ingredient with Y ingredient cause stomach problems?) Someone else handles the *packaging* of the product. (Should it

M. McGrath
Opus International, Inc., Deerfield Beach, FL, USA

R.W. Hartel, C.P. Klawitter (eds.), *Careers in Food Science*,
DOI: 10.1007/978-0-387-77391-9_3, © Springer Science+Business Media, LLC 2008

be packaged in a bag or a box, and why? How does the packaging affect the quality and the shelf life of the product?) And someone else handles the *process development* part of the project. (How should it be cooked? Is it fried, baked, broiled? Why? For how long and at what temperature? What is the cost-effective way of doing all these things to the new snack product?) *Sensory science* is the field of consumer testing. Once a prototype of the product is made, a small group of people (friends, neighbors, other employees, etc.) are gathered to taste the product. Do they like it? What do they like about it? The taste? The smell? The texture of the new product? and why? It is the sensory group, using scientific techniques, who work with the marketing department as well as the research and development department to assure that the consumer will buy the product. (If the consumer thinks the product is too spicy, they won't buy it.) This is a team effort, each member working together to pull the project together and make a safe, tasty snack.

Next door is the *basic research* department. *Basic research* includes developing the *ingredients* that are used to develop the product, such as new starches, thickeners, gums, flavors. How do you make that snack taste spicy? What flavors or ingredients can you use or create that would be best suited for the new product? This group supports the new product development team.

Want to know what ingredients are in the competitors' product? If you like to use sophisticated analytical equipment, then an *analytical chemist* position could be an option. You might be able to make a product better and cheaper.

Now that we have a product that tastes good and is safe to eat, how does the company manufacture it on a large scale? It's one thing to make 50 snack pieces in a pilot plant, now we need to commercialize the product so that hundreds of thousands of these snacks can be made in a day. The *process development* team member determines how to manufacture the product as efficiently and as safely as possible. What equipment should be used? How should the manufacturing line be set up so that the process is streamlined and cost efficient? While the product is being manufactured, the *Quality Assurance* team, both the manager and the technicians, make sure that the food is safely manufactured. Is the equipment clean? How is it cleaned, and with what products? How are the products and ingredients stored? Are the ingredients that are used for the product within the specifications that were requested from the supplier? Are the other production employees following the food safety rules in the plant? Who develops those rules? This is the role of the quality assurance manager and his or her team.

Do I need a graduate degree? That depends on what field you want to pursue. If you want to work at the plant level, that is, a quality assurance or food safety manager in a food manufacturing plant, *generally*, you do not need an advanced degree. If process development is your choice, an undergraduate degree in chemical engineering, food engineering, or agricultural engineering is generally sufficient. If your calling is to be a *research chef*, someone who works in the new product development department of a food/ food ingredient, or food service company (a casual dining or fast food restaurant chain headquarters) then the best approach to this career choice would be to have an undergraduate degree in Food Science, and a 2 year culinary degree from a reputable school such as Culinary Institute of America, Johnson and Wales, etc. Master's degrees are always preferred for all research and development positions. Basic research is the one area that usually requires a PhD.

This does not mean that you are "overqualified" for any of the listed positions if you have a PhD. An advanced degree in the field of food science is almost always preferred, and never punished. However, keep in mind that your advanced degree may make you "unaffordable" to, for example, a smaller company, who cannot afford to pay for the degree. A company seeking a food scientist with only a BS or MS may believe that the job opening itself does not merit a PhD, and may therefore limit their search to those who do not have one. But overall, an advanced degree in food science will give you an advantage over the next candidate who does not have one, and in the long run, you will earn more money in your career by spending an extra few years in college.

No matter what field of food science you prefer, if you chose food science, you will rarely be without a job. Some scientists need to limit themselves to a geographic location for personal reasons. This may make a job search appear more difficult at times, particularly if there are only a few food companies in that area of the country. However, there is such a shortage of food scientists and such a wealth of opportunity in this field, graduates at any level should not be concerned about long-term career growth or opportunity.

To give you an idea of what is required for various types of positions, listed here are typical job descriptions for those with food science degrees.

ENTRY LEVEL FOOD SCIENTIST

A nutritional ingredient manufacturer seeks a Food Scientist to work in their R&D department. The successful applicant will work in a small food science group in the development and applications of novel systems for the delivery of their ingredients into foods. The Food Science Group forms an integral part of a team of R&D scientists that also includes microencapsulation specialists, oil chemists, analytical chemists, biologists, and pilot plant engineers. Applicants should either have a minimum of a BS in Food Science or similar, ideally combined with some industrial experience in food or ingredient R&D. Good organizational skills and the ability to work productively in a team environment are essential attributes. Company would consider those who have recently graduated with a BS or MS if they have had an internship in industry and are residents or citizens. Salary will be commensurate with experience.

DUTIES FOR THE JOB INCLUDE

Food Science support R&D efforts to develop new ingredient delivery forms (coordinate chemical analysis, sensory panels, shelf-life and/or stability studies, reports, and recommendations)

CONDUCTING SHELF-LIFE STUDIES

Sensory analysis of prototypes and commercial products.
Sales support (food product trials and development, reports)
Food product development and research (lab-scale prototypes, plant trials)

GENERAL LABORATORY DUTIES

Providing vital feedback and input to R&D and Production groups on the ingredient performance in food systems

---

PRODUCT DEVELOPMENT TECHNOLOGIST

Take the opportunity to work with one of America's best-loved and well-recognized brand icons. With annual sales exceeding $1 billion, eight manufacturing facilities across the country, and 5,000 employees committed to excellence, Company makes careers rise like fresh bread in a hot oven.

We are the nation's quality and good-taste leader in premium baked goods, from crunchy crackers to decadent cookies, from sweet breads to zesty toast.

Company is seeking an independent professional with passion to join our team in our corporate research center in the Northeast.

As Product Development Technologist, you will play a key role in developing new products from concept through pilot plant to full-scale commercialization. You will lead projects in the areas of new product development, line extensions, product/process improvement, quality improvement, cost reductions, and problem solving.

SKILLS AND EXPERIENCE

- Bachelors degree in food science or related technical discipline preferred, master's degree a plus;
- Minimum 2–3 years successful product development experience in baking, frozen, or snack food industry including food applications, formula management, and specification development;
- Ability to deliver against key objectives and timelines;
- Ability to work well, both independently and within a cross-functional team;
- Strong communication skills, both written and verbal;
- Professional maturity;
- Attention to detail.

SENSORY SCIENTIST

Alcoholic beverage manufacturer seeks a Sensory Scientist to be based at their Southeast location. This individual will design and execute sensory and consumer tests providing information, samples, and details of test executions to partners, analyze sensory and consumer data, and provide recommendations.

Position requires 3 or more years of experience in sensory evaluation, including descriptive, discrimination, and Hedonic testing, BS or MS degree, and strong communication skills. An MS or PhD student would be considered if pursuing sensory science in their graduate program.

Great opportunity to learn on the job and make an impact!

ANALYTICAL CHEMIST

REPORTS TO

Director, New Technologies

PURPOSE STATEMENT

Provide technical leadership for the development and application of new or improved methods to enhance the understanding of food systems and to improve processes. Provide technical support, guidance, and problem solving to R&D, Corporate QA, and manufacturing plants.

MAJOR RESPONSIBILITIES

- Plan and conduct research, which applies basic concepts, theories, and analytical methods to the understanding of food systems;
- Lead or work with cross-functional teams to solve problems related to food product and process challenges;
- Manage analytical laboratory, ensuring timely, accurate data to meet the needs of the business; and ensuring good laboratory practices are enforced;
- Associate management and training;
- Develop, validate, and apply new methods to evaluate product development or processes;
- Establish quality assurance procedures for methods to ensure precision and accuracy of results;
- Prepare technical reports and proposals;
- Maintain awareness of and communicate changes in technical developments with potential impact on the Company's business.

QUALIFICATIONS/ KNOWLEDGE & EDUCATION

- MS or PhD degree in chemistry, biochemistry, physical chemistry, or food science and 2–4 years experience; or BS degree and 4–6 years experience;
- Experience with analytical instrumentation (e.g., GC, HPLC, IR, AA);
- Project management;
- Method development;
- Fats and oils or emulsions expertise.

WHAT TYPICAL DECISIONS ARE WITHIN AUTHORITY?

- Determine best methodology for a particular test;
- Determine and maintain accuracy and precision of results;
- Design experiments and sampling plans;
- Ensure lab instrumentation is maintained and appropriate preventive maintenance and repairs are scheduled;
- Purchasing of lab supplies and noncapital equipment;
- Recommend vendors and best selection for new or replacement instrument purchases

---

RESEARCH CHEF

POSITION PURPOSE

Enhance the profitable growth of Company by providing our customers with culinary expertise and solutions required to expand their menus and the end use of Company's products.

ESSENTIAL FUNCTIONS

- Assist Executive Chef in day-to-day activities. Provide menu analysis, new concepts, and innovative recipes as assigned for key and target customers and other projects utilizing Client's cheese and noncheese products. Demonstrate products and assist with presentations as needed.
- Develop new product strategies with National Account Managers. Purvey products, produce and cheese samples. Interface with Product Innovation as needed.
- Build relationships with key customers, supplier networks, and corporate chefs/new product development teams.
- Research and report new culinary trends through periodicals, continuing education courses, food trade conferences, etc. Share trends research with customers and sales partners.

EDUCATIONAL/EXPERIENCE/SKILLS REQUIRED (MINIMUM LEVEL)

- Associates or Bachelors Degree in Culinary Arts;
- Restaurant/food service cooking experience of 5–10 years;
- A background in food manufacturing and/or teaching would be helpful;
- Extensive knowledge of cooking techniques, product identification, and recipe development;
- In-depth understanding of customer's back kitchen, equipment, operational, and crew issues;
- Ability to purvey groceries and send samples;
- Positive can-do attitude, creative mind, and the flexibility to adapt;
- Strong written and oral communication skills;
- Computer capabilities, including Word, Power Point, Outlook;
- Ability to work well in a team environment;

- Project management skills;
- Self motivated.

This individual should have a demonstrated passion for food and cooking and present oneself in a highly professional manner. Should be skilled at product demonstrations, training, and interfacing with internal and external customers and coworkers. Valid Drivers License and ability to travel by plane, train, and car is required. Travel is 40%.

APPLICATIONS SCIENTIST: FLAVORS

The successful candidate will be responsible for planning, coordinating, and completing customer-driven beverage applications projects. Projects include new product creations, revisions, and matches, which may be shown in customer presentations. Will also write finished product formulas for costing and supervise junior personnel. Extensive internal and external contact required in a fast-paced environment.

MAJOR DUTIES

1. Provide technical expertise in one or more areas of beverage product development such as dairy technology, protein chemistry, gum/emulsifier technology, vitamin/mineral technology, juice technology, or soft drinks.
2. Responsible for completion of major applications projects for key accounts by creating or developing finished products for sampling.
3. Provide technical support to customers and internal departments on beverage applications issues and sales/marketing presentations.
4. Provide input on sensory evaluation panels, product screening, and reports.
5. Provide input on beverage processing of aseptic, retort, hot fill, etc. finished products.

REQUIREMENTS

Bachelor's degree in Food Science, Chemistry, or related subject and a minimum of 3 years experience in flavor applications and/or beverage product development or MS degree and 1 year relevant beverage experience.

NUTRITION SCIENTIST

OUR COMPANY

Living a healthy lifestyle begins with a good understanding of what Nutrition can do for you and your family. Company has an ongoing commitment to provide consumers with a range of high quality food products designed to meet their taste and nutritional preferences.

CURRENT NUTRITION OPPORTUNITIES

We currently have opportunities in our Nutrition Science Group for Nutrition Business Partners and a Director-Nutrition Science.

NUTRITION BUSINESS PARTNER

- Together with the Senior Nutrition Business Partner, we provide nutrition leadership and direction to Company's business strategies and insure the successful execution of the business/marketing plans. In this role, you will be responsible for working with marketing and advertising agencies to identify and develop the nutrition aspects, content, budget of portfolio, and brand strategies, and influence the implementation of such plans. Additionally, you will provide value-added nutrition services for all Company's internal customers (sales, marketing, market research, consumer affairs, food service, public relations, legal etc.), identify opportunities (nutrition, influencers, regulatory) for brands and other business growth initiatives and support/protect/enhance branded nutrition messages and positioning. The Nutrition Business Partner also influences outside health and professional organizations on Company's nutrition strategy.

QUALIFICATIONS

- BS degree in Foods and Nutrition (graduate degree preferred)
- 5 plus years of industry/clinical experience
- Current membership in professional associations such as the American Dietetics Association

Demonstrate excellent verbal and written communication skills as well as negotiation skills, conflict resolution, and problem solving skills. Be able to work in teams and to influence others.

TECHNICAL SERVICES SPECIALIST

Multimillion dollar manufacturer of frozen dairy desserts seeks a Technical Services Scientist II in our Research and Development Department in the Midwest.

This position will provide technical expertise in the areas of product, process, and formulation to our manufacturing facilities. Responsibilities include providing technical expertise to modify and improve formulation. Be responsible for assigning, coordinating, and executing functions necessary to bring projects to completion. Prioritize projects based on company and departmental goals and objectives. Independently work with plant personnel to understand issues, plan resolutions or improvement methods, implement and test plans, evaluate action plan, and standardize or modify plan.

Understand and operate pilot plant and analytical equipment. Possess understanding of Company's formulations and manufacturing processes along with the ability to train others. Be responsible for staying current on food industry technology and innovation for application at Company. Administer direct supervisory responsibility for Technical Services personnel involving human resource functions including selection, hiring, training, performance, evaluations, promotional recommendations, and work schedules.

In addition to the responsibilities listed earlier, other duties may be assigned by your supervisor as dictated by business necessity.

The requirements for this position are Bachelor's Degree in Food/Dairy Science or a related field; graduate degree preferred. Preference will be given to candidates with related experience.

Salary is commensurate with experience.

FOOD MICROBIOLOGIST

JOB DESCRIPTION

Develop and implement innovative solutions in food safety for food service systems.

REPORTS TO

Food Safety Group Manager

RESPONSIBILITIES

- Organize and evaluate new and existing projects for microbiological safety.
- Serve a leadership role within Company food safety microbiologists, and as a liaison with external regulatory, academic, professional, and research foundations.
- Initiate field research studies in the food service industry in the area of microbiology/food safety.
- Manage technical and budgetary aspects of multidiscipline research projects with emphasis on microbiology.
- Design protocols and work plans for systems challenge experiments to be performed in-house or that utilize external resources.
- Provide microbiology/food safety expertise and technical assistance as needed.
- Assess new and existing technologies for innovative applications to food safety in food service.
- Work with project teams to develop creative solutions to food safety issues.

EDUCATION, EXPERIENCE, AND SKILLS REQUIRED

- PhD in Microbiology or a related discipline
- 10 or more years' experience in food safety/microbiology
- Recognized leader in his or her field
- Practical knowledge of food and food processing, as well as factory/implementation experience
- Experience in the following areas is highly desirable: sanitary equipment design, thermal processing, chemical sanitation, and risk assessment
- Excellent written and oral communication skills, as well as project leadership skills are required

---

PLANT QUALITY ASSURANCE MANAGER

Midsize manufacturer of beverage products seeks a quality assurance manager in their Phoenix area plant. This Company's customers are some of the largest and most well-known hotel, restaurant, and food service distribution companies in North America, as well as regional and local favorite restaurant chains, resorts, healthcare facilities, and a variety of entertainment venues.

RESPONSIBILITIES

1. Supervise two lab techs, possibly three techs. Presently there is QA coverage for first and second shift.
2. Develop and analyze statistical data and product specifications to determine present standards and establish proposed quality and consistency expectancy of finished product. Formulates and maintains quality control objectives in accordance with regulatory requirements, corporate policies, and goals and to manage current inspection programs and procedures.
3. Direct the activities of Quality Assurance staff, ensuring policies and procedures are followed.
4. Work with vendors to ensure quality of all purchased components and ingredients.
5. Manage quality control training programs.
6. Communicate issues and other concerns regarding quality to management, plant personnel, customers, etc.
7. Keep abreast of innovations and new methodologies for continuous improvement of QA programs.
8. Follow HACCP, SSOP, and GMP standards.

Company prefers a Bachelor's degree in food science or microbiology. Bilingual capability is strongly preferred. Plant quality assurance manager needs to be a good communicator as he/she will interface with customers, management, vendors, regulatory officials, as well as peers and subordinates.

Salary would be commensurate with experience.

QUALITY AUDITOR

JOB FUNCTION

To assure that Company's products are produced in compliance with pre-established standards and specifications.

DUTIES AND RESPONSIBILITIES

- Develop and update, as needed, the Company's audit form and requirements to encompass Company and customer requirements and regulatory compliance.
- Audit semiannually the Company's manufacturing plants for compliance to our specifications, customer specifications, and regulatory requirements. This will include HACCP, GMPs, and Biosecurity and also includes evaluation of quality control procedures. Participate in copacker audits as needed.
- Responsible for auditing offsite warehouses, Company owned, and others.
- Responsible for inspections and quality reviews of key Company ingredient and packaging suppliers.
- Keep senior management informed as to the condition, risks, and opportunities in all facilities that are audited and reviewed.
- Support the plant quality control managers in evaluating special quality problems.
- Coordinate and issue reports of results from the Company lab split sample testing program. Assist in ongoing Quality Training Programs.

JOB QUALIFICATIONS

- BS in food science or related field.
- A minimum of 5 years experience in dairy processing or quality control.
- Strong attention to detail.
- Ability to work on multiple projects at once.

Position will be home based, anywhere East of Chicago, and north of Maryland.

Travel will be 80%, primarily in the Northeast.

FOOD SAFETY MANAGER

Leads the Food Safety Standards Team in the definition and communication of HACCP and food safety related prerequisite programs. He/she will be responsible for interfacing with Corporate Quality and Operations in development of implementation plan.

Be responsible for maintenance of global hazard analysis structure for select categories of ingredients.

Be responsible for supporting global new product innovation program.

Be responsible for the food safety review and approval of new product development initiatives. Be responsible for establishment of food safety standards for new and existing suppliers. Support Supplier Management in assessment and communication of these standards.

Be responsible for participation on Allergen Steering Committee. Assist committee in evaluation of risk and strategies/options that mitigate risk.

WORK ENVIRONMENT

The incumbent should have knowledge and ability to integrate food law, food technology, food chemistry, microbiology, cereal science, and agricultural practice. The incumbent should possess leadership and communication skills, and diplomacy that will promote good relations with all facets of management and other technical and scientific disciplines in the effective execution of projects to accomplish corporate objectives. This individual must be able to function effectively as a project manager, planner, coordinator, communicator, and scientific resource.

TRAVEL

Approximately 20–40% traveling. Product startups, audits, seminars, plant support related to internal food safety issues.

Mostly United States and North America with occasional global travel

EDUCATION/EXPERIENCE

The ideal candidate will possess a PhD in Microbiology/Food Technology/Food Toxicology/Chemistry/Food Science, or related area with at least

3–5 years industrial experience, or MS and 6 years. Experience must include operations or plant exposure. QA/Food Safety backgrounds if possible.

MANAGER, REGULATORY AFFAIRS

The Manager Regulatory Compliance is responsible for the detailed and current knowledge of domestic and international regulations, and compliance requirements pertaining to flavors and similar food ingredients for the division. This position requires an in-depth knowledge of all the protocols and procedures for raw materials, intermediates, and finished products from the point of view of FDA, USDA, DOT, Customs, and similar international agencies.

MAJOR DUTIES

- Understand, interpret, and apply the current regulations from government regulatory agencies as applicable to flavor business across the division, worldwide.
- Publish and present interpretations and findings from all regulatory meetings and literature in a manner that will drive Company to the forefront of all regulatory compliance activities.
- Develop, implement, and/or maintain programs for labeling, ingredient statements, and nutritional data sheets for Company's products.
- Provide documentation to Regulatory Compliance Coordinators for distribution to customers (natural certification, GMO status, NAFTA certification, etc.)
- Review product information for correctness. Enter and/or update information in mainframe as necessary.
- Provide ingredient statements and nutritional information for customers.
- Provide ingredient statements for specification sheets and international labels.

KNOWLEDGE, SKILLS, AND ABILITIES

This position requires a minimum of a Bachelor's degree in Chemistry, Food Science, or related fields. Three to five years of experience in the area of regulations and compliance is preferred. One to three years in a supervisory capacity is desirable. Knowledge of computer programming is necessary.

ANALYTICAL/INTERPRETIVE COMPLEXITY

The job requires staying current through reading, participation in FEMA, and other technical associations, interaction with agencies, etc.; interpreting information; and, as a result, developing new programs or systems pertinent to complying with all laws and regulations; involves the use of computer programming and algebra for modeling.

PLANNING

Approximately 20%–30% of the time is taken in understanding and assessing of regulations.

What is Ahead for the Food Science Industry

Although we do not have a crystal ball, we see a lot of activity in "new" areas. Consumer trends and global issues, of course, influence the focus of food and food safety. All of these areas create new job opportunities in food science.

Obesity is the number one health issue in the United States. With over 30% of Americans "overweight" or "obese," the consumer is demanding changes in their food products. The *trans* fat issue is a perfect example of the food industry accommodating what the consumer deemed necessary to make products "healthier." More emphasis on nutrition is occurring, particularly for the children and young adults market. The food and food service industry is already moving forward offering "low fat" and lower calorie products. "Good for You" products are becoming more mainstream.

Nutraceuticals, dietary supplements, functional foods, and fortified foods is another field to watch. Foods to protect us from common illnesses, colds, and flu may already be on the market. Health claims are carefully monitored by the FDA, but we have already seen products like cranberry juice given credence to claims of keeping us healthy in a variety of ways. Omega 3 products, probiotics, and soy products have already become common words, yet there will be much more attention to these types of products in the near future. More dollars to study in basic research and new ingredient development, in government, academia, and in industry will be available due to this growing trend.

The Aging of Americans is another area of focus for the food industry. Baby boomers are retiring. However, unlike previous generations, those who are retiring want to stay much more active. New products are already being developed toward this market segment. Gloucosamine and botanicals, for example, are being marketed to help fight inflammatory diseases and maintain joint health. Cardiovascular diseases are being researched by food scientists as well, and food products are being developed to support "heart health."

Children and Diabetes is being researched at length by many groups; doctors, nutritionists, food scientists, etc. This is a critical area, as the numbers of cases of children's diabetes has skyrocketed, mostly due to poor nutrition, diet, and lack of exercise.

Threat of terrorism is changing food safety procedures at all levels. Companies have tightened controls, and even stricter food safety guidelines. There are enemy threats regarding poisoning of water supply or other necessities. Emphasis on protecting our population will be more prevalent.

More food imports from overseas is prompting food and food service companies who purchase food ingredients from overseas to need to revamp their quality control policies and procedures. Currently, there is not enough policing of overseas food plants. As a result, the buyers of those products only know what they have been told, not what they have *observed*. There are questions regarding specifications and safety standards being met. With the current rate of recalls escalating, this will be a major food safety focus. Standards will be tightened and specifications will be strictly enforced.

The Hispanic population explosion is changing product development emphasis. New products are being developed to support this demand. These products will be much more conventional, thus gearing the marketing of the product to attract not just the Hispanic market, but everyone who shops in the grocery store. An example of "perfect" marketing might be the fact that salsa is the number one selling condiment in the United States, overshadowing its predecessor, catsup.

Culinary trends happily are on the rise. It is great to eat healthy foods, but they have to taste good! If a cookie is low fat, and tasteless, it stays on the grocers' shelves. If soy milk is good for you but has a bad aftertaste, no one is going to buy it a second time. Products need flavor, and those companies who know how to add flavor without adding calories will be the winners. Flavor manufacturers already have teams of Research Chefs, or "Culinologists" on staff. Many of the European companies, for example, Nestle, Unilever, have had chefs on staff for years. US food and food service companies are gradually joining this trend and seeking scientists who are "foodies," (people who love food), as well as adding chefs to their product development teams to make their products taste better.

How do these trends affect the job market? They escalate the need to find scientists to develop safe, tasty, "good for you" products. We need scientists to keep us safe; from acts of terrorism to the crazy guy down the street who is on a mission to poison us all. All people need to eat, and everyone wants to have an active and healthy lifestyle. Food scientists play a critical role in these basic fundamentals of life. Think about it, if not for the food scientist, who is responsible for our food supply?

How do I learn about changes in the food industry? There are many resources to keep us up-to-date regarding trends, new products, and new methods.

The Institute of Food Technologists (http://www.ift.org) updates their Web site daily with new information important to all food scientists, such as trends, changes in food companies' focus or leadership, and food safety issues.

The Journal of Food Science, published by the Institute of Food Technologists, is a wealth of technical information regarding current research projects done by academia and government.

Specialty Food News (http://www.specialityfoodnews@foodinstitute.com) has a daily email full of new trends and industry information, geared more toward the non food scientist.

Nations Restaurant News (http://www.nrnnews@nrn.com) is a great resource, both online and their weekly paper, on the food service industry.

And the local newspaper and online news journals always keep us up-to-date about food and food safety issues.

What will the Market Look Like for Food Scientists in the Future?

There is a real shortage of food scientists graduating from US universities. In 2006, less than 500 students in 43 universities offering food science programs graduated with a BS in Food Science (see Table 3.1). About a third of those students continued on to graduate school. Consider the fact that there were about 5,000 chemical engineers who graduated the same year gives you an understanding of how few in numbers are the food scientists. With the "baby boomer" food scientists retiring, there is a dramatic "brain drain." The Institute of Food Technologists as well as many of the universities offering food science programs have tried to support the industry by promoting food science to high school teachers and students in various ways: developing television shows, providing teaching materials, offering courses to teachers, etc. Time will tell if these programs work. For now, according to the research done by the Institute of Food Technologists, we will continue to have a shortage of food scientists until at least 2010.

Table 3.1 Undergraduate Student Survey, 2006 by Opus International, Inc.

University	Number of undergraduate students	Number of graduating seniors	Number of seniors to graduate school
UC Davis	160	23	5
Kansas State University	134	18	5
Purdue University	125	22	5
Pennsylvania State University	100	25	3
Cal Poly	95	26	6
Cornell University	85	20	10
Iowa State	78	18	5
Oregon State	76	16	0
University of Wisconsin	74	20	6
Michigan State	73	12	3
Alabama A&M	72	6	5
Brigham Young University	65	18	5
University of Tenn.	65	13	2
Texas A&M University	61	20	1
University of Nebraska-Lincoln	57	10	4
Ohio State Univerisity	55	16	6
University of Illinois	53	14	6
University of Minnesota	53	10	5
Rutgers	50	10	2
University of Florida	48	17	9
Virginia Tech	45	9	2
University of Georgia	45	17	5
University of Missouri	45	5	2
University of Vermont	45	9	2
Clemson	45	5	2
University of Idaho	43	10	4
North Carolina State University	42	11	3
University of Mass.	38	15	5
University of Arkansas	31	12	3
University of Kentucky	28	6	2
Louisiana State University	28	5	1
Washington State University	23	11	2
Utah State University	22	5	2
University of Maryland	20	3	0
North Dakota State University	20	2	0
University of Delaware	20	3	1
Mississippi State University	18	9	4
San Jose State University	12	2	0
University of Maine	5	0	0
Total in USA	2154	473	133

Summary

So, what is the bottom line? Why should you pursue the field of food science? If you are scientifically inclined, but prefer practical application of science, food science is for you. The food scientist is challenged every day, learning new things, and being part of the huge food industry. They are involved in something that everyone on the planet eats, plants, cooks, etc. every single day, and cannot survive without. If you go to a party and tell your friends that you are pursuing a degree in food science, they will not know what you are talking about. But when you talk about developing new food products, remind them of all the frozen and convenience foods now available in the market place, they will begin to understand. When you mention food safety, they again will be reminded of recalls, *E.coli* issues, and food poisoning. They will see how important your job is. You will have an integral part in protecting people. Career wise, you will be secure. When and if there are layoffs in the food industry, it is rarely in the scientific community. Production workers, sales and marketing, and senior executives are more at risk in a layoff than a food scientist in R&D or QA. Food scientists with a BS degree are very hard to find, and when they have proven themselves with a company, the company will do everything they can to keep them. Salaries are fair. They are twice the level of those with a general science degree such as Chemistry or Biology, and very comparable to those in the pharmaceutical industry. There is upward mobility for scientists, either up a technical ladder or a management ladder. There is opportunity to go into other areas of the industry: technical services, marketing, sales, or a senior leadership role. What else could you ask for? Good money, growth opportunities, plentiful jobs, job security. What else do you need? Go for it!

Part II
The Undergraduate Student Experience

Chapter 4
Making the Most of Your Undergraduate Experience

Christina Klawitter

Preparing to write this chapter on how to make the most of an undergraduate experience, the take-home message from an enrichment retreat I attended as a student came rushing to mind. After a week spent exploring our professional passions and our personal strengths and challenges, the retreat facilitator wrapped up by giving each participant a foot long, brightly colored bungee cord. All the participants began stretching the cords in amusement as he asked a simple, but profound, question: *Over the next year, how will YOU stretch yourself?*

One of the great things about being a college student is that choices and chances to stretch yourself abound. You choose a major to pursue and how you prepare yourself for a career in that field. You choose how often you meet your advisor and whether you approach those meetings as a chore or as a learning opportunity. You choose how to spend your time out of class; in other words, you choose how much time you spend studying and how engaged you become with your university and community. These are just a few of the many choices you will face. You can play it safe or you can stretch yourself with your choices. To make the most of your undergraduate experience, try to make choices that will stretch you out of your comfort zone.

Indeed, a food science curriculum will stretch your academic mind and skills as you develop the content knowledge relevant for professionals in the food science field. Faculty at your educational institution will insist on a high level of technical and scientific competence. When you graduate, your knowledge of food microbiology, chemistry, and engineering as well as nutrition, sensory science, and business related to the food industry will be

---

C. Klawitter
University of Wisconsin, Madison, WI, USA

R.W. Hartel, C.P. Klawitter (eds.), *Careers in Food Science*,
DOI: 10.1007/978-0-387-77391-9_4, © Springer Science+Business Media, LLC 2008

well established. By contrast, you might have to be a bit more intentional in order to stretch yourself and develop what writer/researcher Bill Coplin (2003) calls your "know-how skills." In his book, *10 Things Employers Want You to Learn in College*, Coplin argued, based on data from the National Association of Colleges and Employers (NACE), that grade point average (GPA) does not top the list of characteristics that will compel employers to hire you. Rather, your ability to solve complex problems, work with and influence others, manage conflict, think critically and ask good questions, communicate in writing and orally, work hard, and behave in principled and ethical ways will position you for success far more effectively than your GPA. That is not to say that GPA is not important, but stretching yourself academically, professionally, socially, spiritually, and physically will advantage you when it comes time to take some postgraduation steps. In this chapter, the experiences of Nicole, Tim, and Annie, several exceptional undergraduates who have made the most of their undergraduate experiences, will be woven together to give you some ideas for making the most of yours.

Move Beyond Your Comfort Zone

Making the most of your education requires you to open and expand your mind. It requires you to push yourself outside your comfort zone. It requires you to be open to experiences that might change you. Nicole was not sure what she would face when she decided to study abroad in Chile. It was not required for her major but she saw an opportunity to stretch herself by becoming immersed in a language in which she was not fluent, adapting to a new educational system, and navigating an unfamiliar culture. Like Nicole, many students say that studying abroad is a life-changing event. They meet people unlike those they have met before. Their long-held assumptions about life, family, and ways of doing things are challenged. They discover parts of themselves, such as confidence and an ability to adapt, which they did not even know existed. Now, doesn't this sound like something to experience?

Can't study abroad? No problem, because there are plenty of chances to push yourself outside your comfort zone on your campus. Take a variety of classes outside the topics with which you are familiar and comfortable. Take a class just because it sounds interesting, even if it does not meet a requirement. Try subjects completely outside your major such as language or culture, history, or literature. Attend campus lectures on topics beyond food science. Initiate friendships with people who are different from you in some way. Volunteer in the community with a cultural group different from yours to serve and to learn.

These may not be the right activities for you, but remember that your goal in college is to become educated in the broadest sense of the word. What does being educated mean to you? What do you think being educated means to people who will evaluate you when you graduate? What do you want to learn during your time as an undergraduate? What experiences will push you out of your comfort zone? At the end of college, how do you hope to have grown and what can you do to be sure that growth happens?

Self-Assess and Utilize University Resources to Grow and Learn

Another strategy for stretching yourself is to learn about your strengths and weaknesses. Sometimes, this can be accomplished with some honest self-reflection. Self-assessment tools can also help and can usually be taken in career centers. Another simple idea is to ask for feedback from family, peers and advisors, professors or mentors who know you well. Ask them what you are good at, and ask them to identify skills that you could enhance. Follow that up by asking if they have any specific ideas of methods you could use to improve. Even though you should plan to repeatedly assess your strengths and weaknesses throughout your life, college is an especially good time to do this because there are so many (free) resources to help you grow and learn.

Annie was particularly good at self-assessing and developing a plan to improve herself. She called it her "making lemonade" plan – the process of taking what she was given and finding ways to make the most of it by utilizing resources. She explained that her work ethic, maturity, and general intelligence and ability to learn comprised the best of what she had to offer, but her "lemons" (in Annie's case, her communication skills and confidence) sometimes got in the way. In order to make lemonade out of her lemons, Annie engaged in a communication skill development group that met weekly. She was forced to listen and give others feedback and then had the opportunity to practice her own oral communication skills and receive feedback. She also engaged in a leadership development seminar, which fostered her confidence. Neither of these opportunities was required for graduation; yet, she saw a couple of resources to help her develop and chose to "make lemonade." What are your strengths and how can you further hone them? What are your "lemons" and what resources exist on your campus to help you "make lemonade?"

Immerse Yourself in Activities that Foster Your Professional Development

Finding ways to foster your career interests and gain professional skills is essential to making the most of your education. Tim engaged in several internships, each one a little different so he could explore various industries and positions within those industries. He attended all the career fairs, resume writing workshops, etiquette dinners, and mock interviews. He went to company information sessions, even when he was not sure it was a company of interest to him. He asked a lot of questions. He developed his professional network by following up on every lead recommended to him, even if he was not sure it would lead somewhere productive. He regularly reflected on the relationship of his undergraduate experiences to his future career plans. Instead of focusing on doing things solely for resume building purposes, he focused on developing his professional passions and skills. His commitment to learning led to an impressive resume and a variety of job offers upon graduation. What can you do this semester and this year to foster your professional development?

Engage in Meaningful Conversation with Faculty

Research shows that students who are most satisfied with their college experience have regular and meaningful interaction with faculty (Light, 2001). If you are wondering why this might be, consider my observations about students who are connected to faculty on their campuses.

Students who build connections with faculty are likely to:

- seek and receive good academic advice
- be mentored to pursue careers that are a good fit for them
- consider graduate school and have the confidence that they could succeed there
- engage in mentored research projects as undergraduates
- obtain strong recommendation letters
- solve academic and personal problems with minimal setbacks to their degree progress
- feel connected to their departments and universities
- be more engaged in out-of-class learning activities, such as student organizations, study abroad, leadership programs, and so forth

Make it your goal to know at least one faculty member before the end of your freshman year. Developing a faculty connection can help a big university seem smaller or a small university to seem even more personal.

You are likely to find that faculty are very interested in helping you and getting to know you, so be bold and seek those connections early.

Develop Your Leadership Skills

Nicole, Annie, and Tim went to great lengths to develop their leadership potential by participating in enrichment opportunities of all kinds, including leadership retreats and courses, and workshops and conferences. They took leadership roles in student organizations, in projects at their internships, and in interpersonal interactions in their research labs. Developing leadership potential is important for a couple of reasons. First, as an educated person, you have a responsibility to make a difference, in your community, in your profession, at your university. You *will be* called upon to lead, somewhere, sometime, and you need to be ready to be effective. You do not have to move mountains or lead a crusade, but you should think, for example, about what you can do to leave your university better than when you arrived. Also, developing your leadership potential is important because research shows that leadership skills can be the tipping point when an employer is faced with a hiring decision between two comparably qualified candidates (NACE, 2006). In other words, given two people with equal attributes, a student with leadership experience is more likely to get the job.

The term "leadership" can be loaded with historical images of positional leaders with great authority demanding action from their followers. Challenge yourself to consider alternative views of leadership. For example, consider Badarraco's (2002) claim that great leaders are not public heroes or high-profile champions of causes. Rather, he argued that great leaders move patiently and persistently, doing the right thing for themselves and their organizations, day in and day out. Or consider the notion that leadership is really about relationships through which groups of people accomplish change or make a difference to benefit a greater good (Komives et al., 2006).

Consider the skills that effective leadership, given these perspectives, might require: self-awareness and a sense of what is ethical, solid communication skills, an ability to involve and relate to others, an ability to recognize important issues, and develop a plan to engage others and make positive change. How could you develop some of these skills?

1. Read about leadership and the requisite skills. An easy way to stimulate some initial thought would be to read Komives, Lucas, and McMahon's *Exploring Leadership: For College Students Who Want to Make a Difference*. There is a host of books and articles written about the topic of leadership—visit a bookstore's leadership section or search the Internet and see what topics jump out at you.

2. Practice leadership. For example, engage in student organizations and offer to lead a service or social project. Rally a group of people to volunteer in the community for a cause you care about. Offer to be the group coordinator for a team project in one of your courses. Suggest improvements for how your research team communicates with each other.
3. Reflect on what you did. Keep a journal or ask someone you trust to talk about your experiences with you.

Summary

Even if the stories of Nicole, Annie, or Tim are not convincing or the advice in this chapter are not persuasive, then at least consider one of the main conclusions made by Pascarella and Terenzini (2005) after more than 20 years of research on college student development. They concluded that the impact that college has on students depends almost entirely on students' effort and involvement in the academic, interpersonal, and extracurricular activities available to them. Have you ever heard the phrase "You have to make it happen?" Pascarella and Terenzini's research suggests that this phrase applies to the undergraduate education; not surprisingly, you have to be an active participant in order to get the most from your education.

Remember the bungee cord that prompted me to stretch myself? Even though I am no longer an undergraduate student, that bungee cord still hangs on my bulletin board. It serves as a regular reminder of the lesson that I am indeed responsible for stretching myself, for making lemonade out of the lemons I have been given. So the question looms, what will *you* do over the next year to stretch yourself? Realizing that you have a great deal of control over the answer to that question is the first step in making the most out of your undergraduate experience.

References

Badaracco, J. L., Jr. (2002). *Leading Quietly: An Unorthodox Guide to Doing the Right Thing*. Harvard Business School: Boston, MA.

Coplin, B. (2003). *10 Things employers want you to learn in college*. Ten Speed Press: Berkeley, CA.

Komives, S. R., Lucas, N., and McMahon, T. R. (2006). *Exploring leadership: For college students who want to make a difference*. Jossey-Bass: San Francisco, CA.

Light, R. J. (2001). *Making the most of college: Students speak their minds*. Harvard University Press: Cambridge, MA.

NACE. (2006). Job Outlook 2007. NACE: Bethlehem, PA.

Pascarella, E. T. and Terenzini, P. T. (2005). *How college affects students*. Jossey-Bass: San Francisco, CA.

Chapter 5
Campus Career Services and Centers

Jennifer Neef

The first step toward a successful professional career path should be making a conscious decision to pursue a specific career. Based upon the Cognitive Information Processing approach to career development and services, Reardon et al. (2006) has grouped the process of making career decisions into three domains: Knowledge Domain, Decision Skills Domain, and Executive Processing Domain. The Knowledge Domain includes assessment of one's own skills, values, interests and employment preferences, and gathering occupational information. This is commonly known as career exploration. The second domain is the process of making a decision. It is within the Decision Making Domain that one's career goals become more sharply focused, based upon knowledge gained in the first domain. The third domain, Executive Processing, is thinking about the decision that has been made—both positively and negatively—and acting on it. The third domain includes professional/academic preparation, experiential learning, and the job search. Students should find career counseling services and resources that meet their needs at any point in the career decision making process—from exploration through offer acceptance and negotiation.

Career Services on Campus

The way in which career services are delivered to students on campus is typically structured in one of two ways—a centralized system where services are provided from a single campus office or a decentralized system in which services designed for specific groups of students are provided from various offices on campus. Centralized offices tend to be supported and administered

---

J. Neef
University of Illinois, Urbana, IL, USA

R.W. Hartel, C.P. Klawitter (eds.), *Careers in Food Science*,
DOI: 10.1007/978-0-387-77391-9_5,

by a single campus entity, such as Student Affairs, and are designed to serve all students, regardless of academic discipline. In the decentralized model, services tend to be tied to an academic discipline. Students will find services and resources provided within their college or academic department. In a decentralized model, students may find that using services and resources beyond their "home" office are also available and beneficial.

The initial phase of making a career decision is exploration. However, it is not necessarily exploration of jobs, occupations, or career paths. Rather, effective and sound career decision making begins with gaining an understanding of one's interests and values. Selecting a major based upon one's interests and values, as well as acquiring essential competencies for a chosen profession, translates to being prepared for a meaningful and rewarding career and confidence in one's ability to perform tasks pertaining to his/her occupation.

Students may seek the assistance of a career counselor to ensure their major will lead to their chosen career or to determine what career opportunities are available for particular majors. The Strong Interest Inventory® (SII) is one of the most often used tools to assess career interests and determine potential occupations that match those interests. The foundation of SII® is Holland's (1973) RIASEC typology. Holland's theory classifies people into six personality types—realistic, investigative, artistic, social, enterprising, and conventional (RIASEC)—and is widely used in the field of career counseling. The Myers–Briggs Type Indicator® (MBTI) is another instrument sometimes used by career counselors. MBTI® is a long-used instrument designed to determine personality preferences. It does not tie specific personality traits to occupations; however, it can assist students to think about jobs and work environments in which they will thrive.

While these and other similar assessments can be found on the Web and accessed by anyone for a fee, campus career services offices usually offer them to students at no charge or for a nominal fee. Additionally, by accessing these tools through the career services office, students benefit from having a career counselor interpret the results and help them develop an action plan based on the conclusions, which should lead to a sound career decision.

Identifying one's personal values as they pertain to a career, referred to as work values, is also an important part of making a career decision. Work values are related to how one feels about his/her job. People usually feel satisfied and successful when their job matches their work values. Work values relate directly to the tasks associated with a specific occupation and the conditions or settings that are inherent with that occupation, such as work environment, salary, upward mobility, geographic location, work schedule, service to others, contribution to society, and recognition. For instance, one may choose a career in product development because he/she values a research-based job that utilizes his/her creative ability. To be satisfied with this choice, one must also accept that he/she will typically work indoors in a lab setting with regular hours, likely be required to have an advanced degree, and need persistence to achieve results. Determining one's values requires much thought and self-assessment. Career counselors can provide lists of common work values that students can prioritize. Having a prioritized list of work values is helpful when considering potential careers.

It is not uncommon that students' values are similar to those of their parents. Conflict can sometimes arise if a chosen major or career path is not valued by students' parents. Career counselors are able to provide students with information and tools that will help them attempt to resolve this conflict of work values.

Skills and Competencies

Gaining an awareness of one's skills and competencies, or lack thereof, and those that are required for a specific career is also important. Students identifying gaps in their skill set as freshmen or sophomores have more time to take actions that will develop and improve competencies that are relevant to their career goals.

Required core technical competencies for food science majors, as defined by the Institute of Food Technologists (IFT), are grouped into four areas:

- Food chemistry and analysis
- Food safety and microbiology
- Food processing and engineering
- Applied food science

Students can expect to build these core technical competencies by completing their degree in food science, as they are a fundamental part of their academic program. Proficiency in these core competencies can be measured by academic performance in food science courses designed to teach these technical skills. While completing a degree in food science should provide students with adequate technical competencies, students who excel academically and/or those who seek opportunities beyond the classroom to enhance their core technical skills will be highly sought after.

In addition to technical competencies, IFT also recommends that students develop core success competencies. These are commonly called "soft" skills. IFT's list of essential success or soft skills is as follows:

- Communication skills, which are defined as oral and written ability, listening, interviewing, etc.

- Critical thinking/problem solving skills, which are defined as creativity, common sense, resourcefulness, scientific reasoning, analytical thinking, etc.
- Professionalism skills, which are defined as being ethical, having integrity, respectful of diversity
- Life-long learning skills
- Interaction skills, which are defined as teamwork, mentoring, leadership, networking, interpersonal skills, etc.
- Information acquisition skills, which are defined as written and electronic searches, databases, Internet, etc.
- Organizational skills, which are defined as time management, project management, etc.

The success skills identified by IFT are similar to those identified by employers in the annual National Association of Colleges and Employers (NACE, 2007) Job Outlook Survey. The top skills and attributes that ideal candidates possess as indicated by the NACE Survey are as follows: communication skills, honesty/integrity, interpersonal skills, motivation/initiative, strong work ethic, teamwork skills, computer skills, analytical skills, flexibility/adaptability, and detail oriented.

Like other employers in technical fields, food industry employers assume that graduates with a degree in food science have technical skills and knowledge required to perform tasks that will be assigned or that they can be taught in an efficient and timely manner with on-the-job training. Therefore, after meeting initial educational qualifications for a position, hiring decisions are based upon evidence of soft skills that are required to be successful in the world of work. Soft skills can be developed both in and out of the classroom. Campus-based professional and social organizations, student clubs, and part-time employment provide avenues for students to enhance their soft skills. Career counselors can direct students to specific organizations or campus activities that are beneficial. They should also be able to provide knowledge of how to source a campus-based part-time job. Be aware that campus-based jobs, especially those that require specific technical knowledge, may filled by "word-of-mouth" and not be publicized in a central location.

Exploring Careers

After understanding one's interests, values, and skills, students are ready to begin exploring occupations that match their unique attributes and values. Students can begin narrowing their options by gathering general information about specific occupations. O*NET Online is especially useful as a career exploration tool. It is an interactive database that allows users to search specific occupations or to find occupations based upon their interests, values, and skills. Search results also include a list of related occupations, which enable quick analysis and comparison of many occupations. Another useful online resource is the Occupation Outlook Handbook (OOH). It, too, is an easy-to-use, searchable, online database that provides information about hundreds of occupations. Both O*Net Online and the OOH provide users with job descriptions, earnings information, education requirements, employment outlook, and more. Both resources are regularly updated and are provided as a service of the U.S. Department of Labor.

With a narrowed list of potential careers, students should seek more in-depth occupation information. There is no better information than that which comes from someone performing an occupation of interest. Job shadowing programs, externships, and informational interviews provide students a realistic snapshot of what a particular occupation is really like. Attending employer information nights and career fairs are also great ways to learn about occupations and career paths with specific employers.

Job shadows and externships are usually one-day– to one-week–long programs that pair students with a professional in their field of interest. Students "tag along" with whomever they have been paired to learn about the day-to-day responsibilities of a job. These also provide great networking opportunities as students will likely interact with various people throughout the duration of the program. Sometimes, more formal work experiences, such as internships, are derived from job shadowing or externship programs.

If a campus-sponsored job shadow or externship program is not provided, students can create these opportunities for themselves by arranging an informational interview. An informational interview is an information gathering session—not a job-seeking session. Students arrange an appointment with a professional in their field of interest and ask questions about the profession. The informational interview itself is not long—typically 30–60 min. However, considerable effort should be put into identifying the right professional, developing a list of questions, arranging and confirming details of date, time, and location the meeting, etc. Career services professionals can help students prepare for an informational interview and identify professionals to interview. Alumni and contacts garnered through a professional organization, such as IFT, are usually great candidates for an informational interview.

The Job Search

When the time comes to search for the "right" job opportunity, many students find that the search is time consuming and requires perseverance. Everyone's job search is unique and therefore, search tools may be very different from person to person. It is important to identify the resources best suited for the position that is being sought. Time and energy should be focused on using those that are the most effective and efficient.

Students tend to immediately turn to traditional on-campus recruiting (OCR) activities to start their search without considering if OCR is the best means to use for their unique job search. The most widely publicized OCR event is likely to be the career fair; however, there are events beyond the career fair that bring employers to campus, including informational meetings, career workshops, and class presentations.

The types of employers that participate in OCR events tend to be large corporations that devote significant time, financial resources, and human capital to recruitment efforts. The amount of resources devoted to OCR is often related to the number of hires, both intern and full-time, that are targeted. Smaller employers within close geographic proximity to campus or those with strong allegiance to the institution will also likely be involved in OCR.

The campus presence of some employers may only be that of a job posting. They may not be a physical presence. Therefore, it is important that students understand the job posting system on campus. Is there a single campus-wide job board or is accessing more than one board required to see all the positions of interest? The answer to this question may be dependent upon the organizational structure of career services on campus—centralized or decentralized. Whatever job boards exist on campus, one can be certain that those position announcements only represent a fraction of opportunities that are available. Career services professionals can recommend job search resources and techniques that meet the needs of individual students.

When the "right" job is found, it is imperative that students can articulate their qualifications and relevant competencies with written words on a resume (and sometimes a cover letter) or spoken words in an interview. Career services offices provide resources to ensure that students can successfully communicate their competencies. Resume workshops are designed to teach students the basics of resume writing. Students who have never written a resume will benefit immensely from attending a group resume workshop. Attending these workshops will likely frustrate students who understand the basics of resume writing and have already written some version of a professional resume. Their needs cannot be effectively met in a group workshop setting. They are ready for a one-on-one resume critique.

Resume and cover letter critiques are an intense evaluation of what is written. They are more than a simple proofreading exercise. Career services professionals will ensure that the written words accurately and efficiently describe experiences and that those experiences are effectively translated into competencies that are relevant to the position that is being sought. Counselors seek clarification and details about experiences—both listed and those not listed—to ensure students are showcasing their most relevant experiences, skills, and abilities. Students also have the opportunity to ask specific questions that cannot be appropriately or adequately answered in a group workshop setting.

In addition to a critique by a career services professional, students should have their resume reviewed by several other people—friends, parents, advisors, and faculty members. Sometimes, career services offices provide opportunities to students to have their resumes reviewed by industry recruiters. This is an opportunity that should not be missed. Garnering advice from a recruiter involved in the hiring process is invaluable. After several resume reviews, students should consider the comments of each reviewer and then decide if changes are needed. Remember, resumes are a representation of one's unique qualifications and competencies—not a representation of the reviewers' opinions. Also, before paying for a resume review by an "expert," wisely consider what is received for the money. Is the paid "expert" going to provide any insight beyond that from career services personnel on campus?

Sometimes students who understand the basics of resume writing struggle to write a resume because they believe they have no experiences worthy of being listed on a resume. In this situation, an appointment with a career counselor will be beneficial. A counselor can help students find meaning in what they believe are meaningless experiences.

Articulating competencies in the interview is essential. Many career services offices offer mock interviews – take advantage of this opportunity if it exists on campus. Mock interviews that are taped and allow students to view the mock interview are especially beneficial. Seeing and hearing oneself answer questions in an interview setting can be unsettling. However, it is very evident if improvements are needed.

The complaint most often heard from students about mock interviews is that the questions are not related to or relevant for a specific occupation or career within food science. However, questions designed to determine competency in technical areas are not likely to be asked in an interview, and certainly, technical competencies are not the only ones evaluated in the interview. Employers are seeking evidence of "soft skills" during an interview. Therefore, mock interview questions are designed to help students provide evidence of soft skills—communication, problem-solving, leadership, teamwork, etc. Questions centered on those competencies can be asked by anyone, even by someone who lacks a food science background or knowledge about the industry.

Students are strongly encouraged to participate in an etiquette dinner if given the opportunity. Etiquette dinners teach the basic rules of eating; however, the purpose is not for students to learn every rule of formal dining etiquette. In fact, few people remember or follow all dining etiquette rules, not many will notice a knife resting the wrong way on the plate. Etiquette dinners introduce students to different eating styles—formal, American,

European. Students with a basic understanding of the rules of eating are less likely to commit grievous errors that are impolite or distracting to others at the table.

Having acceptable dining etiquette is required of all persons pursing a professional career. In fact, most find themselves being interviewed over a meal at some point in their career. Being familiar with dining etiquette and comfortable at the table will be especially important during stressful interviewing situations.

In addition to mock interviews and etiquette dinners, career services offices provide other interview preparation resources such as:

- Dressing for an interview
- Sample interview questions
- Sample questions to ask the interviewer
- Behavioral-based interviews and answering questions in the STAR format
- Case interviews
- Second-round interviews or onsite visits
- Handling illegal questions

Summary

In conclusion, students are responsible for making their own career decisions. It is not the career counselor's job to make a decision—rather, career counselors provide information and direct students to resources best suited for their unique needs so that a sound career decision can be made. Sound career goals are based upon evaluation of one's interests, values, and skills as well as exploration of occupations and career paths. Once a career has been chosen, students must acquire the educational qualifications and core competencies required to be successful in their chosen profession. Finally, students must be able to effectively communicate their qualifications and competencies on a resume and in the interview. There is a plethora of resources to help students at every point in the process. Students should take advantage of them.

References

NACE Job Outlook Student Version. 2007. National Association of Colleges and Employers.

Reardon, R. C., Lenz, J. G., Sampson, Jr., J. P., Peterson, G. W. (2006). *Career Development and Planning: A Comprehensive Approach, 2 nd Edition,* Mason, OH: Thomson Custom Solutions

Holland, J. L. 1973. *Making Vocational Choices: A Theory of Careers*. Englewood Cliffs, NJ: Prentice Hall

Chapter 6
Freshman Experience

Becky Kuehn

First, coming to college as a freshman, is one of the most exciting times. It is a chance to start over, make new friends, and be independent. But, besides all of the fun and excitement, there is school. The whole reason for going to college is to get an education. Freshman year of college is the time to start thinking seriously about the future. The pressure is on to decide what to do with a college education and this means selecting a major. Choosing a major can be a daunting task since there are countless possibilities.

Is Food Science For You?

However, even having a slight interest in food science serves as a good starting point for freshmen to consider a career in food science. Conversely, there are incoming freshmen who know right away that they want to major in food science, they know exactly what classes to take, and where they want to work and live after graduating. These examples of college freshmen can be divided up into two groups—those who have a vague idea of what they want to do and those who know precisely what they want to do.

Most freshmen students fall into the first category; they have a vague idea of possible majors, and are considering food science as an option, but do not quite know what food science is all about. They could be interested in anything from business to physical therapy, and even food science. The good thing is that there are nearly endless amounts of resources available to college students to help them learn about majors, including food science, and to help them eventually choose an education path that best fits them.

B. Kuehn
University of Wisconsin, Madison, WI, USA

R.W. Hartel, C.P. Klawitter (eds.), *Careers in Food Science*,
DOI: 10.1007/978-0-387-77391-9_6, © Springer Science+Business Media, LLC 2008

Exploring Majors

Exploring majors other than food science is important before declaring a major. Even the freshmen who know that they definitely want to engage in a food science major should consider investigating other fields. Researching other options will ensure that pursuing a major in food science is the right choice and will be exciting and interesting both during college and after graduation.

Information on all college majors is generally easy to find. The first and most accessible resource is the University Web site. Every student should become familiar with the University's Web site and all of the resources that are available through it. The Web site provides an overview of each major, a list of required classes, and contact information for faculty within the department.

Talking with professors, advisors, and the department chair can help a freshman set up a job shadow in a profession of interest. A job shadow is a great opportunity if seriously considering a career in food science. Since there is such a diverse range of careers available to food science majors, spending the time with an actual food scientist provides one way to determine specific areas of interest within food science. A job shadow can also help with networking in the future.

Another good resource is a Majors Fair, where representatives from all majors that the university offers are present. Not all universities offer majors fairs, but if there is one, definitely take advantage of it. Students can walk around the fair, visit the booths they are interested in, and pick up information. The booths usually have a faculty member and older students, so it is another good opportunity to ask questions. Some key points to think about when exploring other majors are what types of classes are required, how many years of schooling are necessary, and what jobs are available.

Freshman year is such a good opportunity for exploring other majors. Just by taking introductory courses, freshmen are able to determine if that major is what they want. For example, a food science freshman will usually take introductory science courses. These courses allow students to determine how interested they really are in the sciences and laboratory work. It is crucial to enjoy your classes. In Food Science, for example, the typical freshman will take introductory courses in biology, chemistry, and physics, along with mathematics, communications, and humanities courses. It is necessary to have a strong interest in science and to be successful in the science courses. Even though it may not be obvious, the knowledge gained from introductory courses is used in classes specific to food science majors that are taken junior and senior year.

Getting Information About Food Science

Since so many food science careers are available, it can be intimidating for students to decide what interests them. However, in addition to a job shadow, the resources available on campus are numerous and easily accessible.

In the food science department at the University of Wisconsin (UW), for instance, there is a one credit Discovering Food Science course. Most other universities will probably offer an introductory food science course similar to the one at UW. This class of around 40 students, who are current food science majors or who are interested in food science, is taught in discussion form. Groups of eight students are led by a senior food science student (who is usually also involved with some aspect of the food science club). Within the groups, topics involving food science are discussed and then one member of the group will present the thoughts of the group to the entire class. The topics covered in class include discussion of current food issues, nutrition labels and claims, and different career paths available within food science. This class is not difficult. It is designed to be fun, every class includes eating foods that are integrated into the theme of the class and one of the activities is making ice cream. Besides being a fun class, "Discovering Food Science" is informative, it makes freshmen think about the future and listen to what seniors have to say about classes, internships, and what they plan on doing upon graduation.

The best place to get information and advice on a one-on-one basis is from an advisor. Talking to an advisor in the food science department is really important, especially as a freshman. It will help with applying for jobs, internships, and scholarships. Freshmen should not be afraid to get to actually know their advisor and discuss and get advice on more than just what classes to take. If a student, even a freshman, shows an interest and willingness to obtain an internship for the summer, an advisor will recommend the student for the job. After that, it is the student's job to submit a resume and a cover letter and prepare for an interview.

Another reason to get to know professors in food science is that they hire students to work in their labs. This has multiple benefits because a student is getting to know a professor and also getting lab and research experience.

The department chair is another important person to get to know and who can also answer any questions. It is especially easy to approach the chair in a small department like food science, so take advantage of this luxury.

Contacting faculty through email and/or in person is extremely beneficial and offers a good opportunity to ask questions. Talking with faculty members is an enormously valuable resource throughout college, so it is good to start communicating with professors, advisors, and the dean, as soon as possible.

Other faculty that can provide information within the College of Agriculture and Life Sciences (CALS) are the Dean or Associate Dean of CALS, and especially the people in CALS Career Services. The people who work in Career Services are always excited when a freshman student comes in. It shows that the student is thinking about the future. Career Services assists students with creating resumes, preparing for and scheduling interviews, and organizes etiquette dinners and career fairs.

Career fairs (which are different from majors fair) are excellent opportunities to find out information about the companies that hire food science majors. Students can hand out resumes and speak with representatives from a variety of companies and schedule interviews. Freshmen can even benefit from attending career fairs and possibly be asked to interview for an internship. Every interview is beneficial and will only improve interview skills and increase confidence. Practicing and preparing for interviews will help in the future when looking for a job or internship.

If you missed the career fair as a freshman, there are still other opportunities to get in contact with companies, and that is at informational sessions held on campus by different companies. Most universities have info sessions, which are usually advertised through the food science club. They are generally about an hour long and include a dinner of usually pizza or subs, a presentation by food scientists from the company, and then a question-and-answer session. The presenters focus on getting across the feel of the company and what its values are, following with a focus on the products, and finally with information on the careers and internships available to food science majors at the company. After the presentation, the company representatives will stay and this is an opportunity to introduce oneself and ask questions. It could be that they might be interested in a freshman for a summer intern position, or maybe they will keep you in mind for the future.

A summer internship after finishing freshman year is a big deal. The purpose of an internship for a freshman is to provide an understanding of the food industry and to build a foundation for future jobs and internships. The internship probably will not be very in-depth or allow for much independence; however, this allows for a wide variety of food science–related jobs within a company to be explored. Figuring out what aspect of food science is most interesting is the main point of a freshman internship. It narrows down interests in order to focus on one career aspect of food science. For example, maybe a student thought he wanted to work in quality control, but after spending a summer working in quality, he found out it was not really for him and would rather pursue some other aspect of food science.

Resources/Opportunities

One of the best resources for exploring food science and the careers within the major is the Food Science Club. The club is student-run with a faculty advisor. Club meetings at UW are held monthly. The meetings begin with an update on club activities followed by a presentation by a company along with a dinner of sub sandwiches or pizza. In addition to the monthly meetings, Food Science Club has volunteer and social activities, and is in charge of organizing company info sessions as mentioned previously.

Food Science Club also offers the opportunity for leadership experience. It does not require a huge amount of time; it is fun, and fits in with a food science major. Employers look for leadership experience on a resume because it shows responsibility, communication, and management skills.

Also through the UW Food Science Club, there are opportunities to get involved with the product development team, College Bowl, and dairy products judging. Different universities will offer different involvement opportunities, but these are the three main ones: The product development team works together to create and develop an original and exciting new food product, and then enters the product into a national competition such as the Institute of Food Technologists (IFT) or the National Almond Board Competition. Participating in product development presents an opportunity to go through the process of developing a new product, including formulating, testing, packaging, and marketing. College Bowl is sponsored by the IFT and is a food trivia competition between schools. Dairy Product Judging is sensory evaluation of dairy products like cheese, yogurt, milk, and ice cream. Both College Bowl and Dairy Product Judging begin with regional competitions in which the winning school continues on to the national competition where the first place winners receive awards. In freshmen year, it can be difficult to get a large role in any of the teams, but it is still beneficial to get involved and experience how the different teams work. Getting involved as a freshman will help you be successful in the future.

Staying Focused

Keep in mind that everything done as a freshman is preparation for the future. The habits and interests developed during freshman year set the stage for the rest of college. It can be difficult to stay focused with so many other distractions in college, so finding that level of balance between academic and social activities is essential to being successful.

For most freshmen, it is the first time away from home and being responsible for studying and earning good grades. Going to lecture, discussion sections, office hours, and review sessions help to keep focus and be successful academically. Finding a study group is also a great way to meet other people and get in study time. Getting to know the libraries and tutoring resources offered by the university are also beneficial. Everyone has different study habits, so find what works and gets you good grades, and stick with it.

Furthermore, in order to be successful academically, freshmen must learn to prioritize. Spending time with friends, studying, and other activities take time. Getting involved is a good thing as long as you are not spreading yourself too thin. Decide what is most important and stick with that. For example, instead of being involved in dairy judging, college bowl, and product development, pick one of the activities and sign up for a leadership role. It is better to be in one student organization and be a leader than to be an inactive member of five different organizations.

With so much to do, freshmen can easily get stressed out by the pressures of college, so it is also important to figure out a good way to relieve stress whether it is by exercising, watching college sports, or hanging out with friends. Also, it is important to get a decent amount of sleep every night, although it can be difficult since there are so many other things to do, and living in the dorms does not help since there are always people around to talk to or watch movies with.

Freshmen year of college is so much fun, experiencing dorm life, meeting so many new and different people, and being independent. It is important to make the most of the freshman experience by getting involved and using the resources available.

My Experience

My interest in a career in food science began in my junior year of high school, but I have always liked to cook and bake with my Mom since I was a little girl. When I got older, I started to experiment with recipes, and then, I took chemistry as a junior and had a really good experience. I proceeded to take AP Chemistry senior year. I had the same teacher for both chemistry classes, and she was phenomenal. After the AP exam, she invited a recent food science graduate from the UW to give a lecture to our class about food science. This visit from a recent food science grad, and especially one from UW, made me even more interested in pursuing a career in food science. Furthermore, I did a project in one of my high school classes that required me to do a job shadow. I shadowed a food scientist who specializes in sensory science, and thought it was such an amazing job. After these experiences in high school, I knew I wanted to be a food scientist. For me, the things that I did in high school were extremely beneficial to me when I came to the University of Wisconsin.

Before actually coming to the University, freshmen generally are required to attend Student Orientation, Advising, and Registration (SOAR). SOAR is a weekend during the summer before your freshman year when you get to spend the night in the dorms, tour the campus, meet other freshman, and sign up for your fall semester classes (most universities have a program similar to this). This orientation session was my first opportunity to show my interest in food science at the University. It is when you can officially declare a major and also pick classes that are required for that major. Besides picking classes, SOAR creates a very social environment that encourages you to meet other students. Most of my friends now at UW have said that they had an awesome time at SOAR and are still friends with people they met during that weekend.

After actually moving into the dorms, school does not start right away. At UW there is what they call "Welcome Week." This is the week before classes when there are endless activities in the dorms for freshman to get to meet each other. There are also welcome sessions at each of the different schools within the university. In the CALS, there was a speech by the dean, followed by an ice cream social, and then we broke up into groups based on majors. This was really a good opportunity for me to meet some of the people in the food science department and even other freshmen in food science. It is extremely important to get to know the other students because those are whom you will be having classes with and who you will be studying with in the future.

Once the semester actually started, I started to have my doubts about food science. At one point during my first semester, I was seriously considering switching majors. I was actively looking into majoring in physical therapy, which has a very different curriculum from food science and also requires additional years of schooling. After looking into physical therapy, I decided to stay with food science. I still sometimes think about other possible majors and careers, but I know that I want to stay with food science. I realized that there are so many different opportunities in food science that if I don't like one job, there is some other job out there that I can do using my degree.

Another huge part of going to college for me was the fact that I would be living on my own, without my parents, and with another girl whom I had never met. Living without my parents was hard for me at times since I was close to them. My parents only lived about an hour and a half away from Madison, so they could easily come up for football games or just to visit for the day. Also, my brother is 2 years older than me and also goes to the University of Wisconsin. If I needed to, I could always go to him for help or even if I just needed someone to go out to eat with. I was really lucky to have him there for me my freshman year. It helps to know or at least get to know older students, especially within your major.

As for living in the dorms with someone I had never met before, I honestly had no idea what to expect. When I found out who my roommate was, I called her to find out who she was and what she was into and figure out who should bring what. This first phone call did not turn out quite how I expected. After living with her for awhile, I figured out that she was a quiet person and kept to herself. This was kind of hard for me because I like to be social and go and do things, but I became friends with other people in my dorm that were just a door away or down the hall. All I had to do was go out on my own and meet other people on my floor. It turned out to be a lot of fun living in the dorms. There are always people around to hang out with, so it is never boring. Now, it is my sophomore year and I live off-campus, and I realize how much I miss living in the dorms and seeing my friends everyday. Of course, there are things I don't miss, for example, having to share a tiny room.

Academically, my freshman year was kind of rough. I remember my first college exam was in General Chemistry II. I studied hard for the exam and thought I knew the material and would do well. However, when I got the exam back, I found out that I got a C on it and I freaked out. As someone who is very hard on myself and expects perfection, I was not used to getting Cs. I did not want to continue to get these types of grades, so I had to change something. I started to go to office hours with my professor and teaching assistant (T.A.). I made sure I would do better on the next exam, which I aced! Anyway, I found out for myself what I had to do in order to succeed in that class. You have to take the initiative and do what is necessary to be successful. Freshman year is the best time to figure out what works for you and what you need to do in order to do well in school. Don't wait until it is too late. This is all part of taking responsibility for your schooling.

Along with the studying and expectations of good grades and involvement in extracurricular comes stress. It is super important to learn how to constructively deal with stress, and I don't mean going and partying all the time. For me, it was running. Running is something that I am passionate about, but also allows me to relax. Going for a long run gives me some time by myself to relax and just think about things, or even forget about things for a little while.

If running isn't your thing, another way to deal with stress that I found to be really helpful was to just talk to your friends in the dorms. Don't be afraid to tell them what is going on in your life and admit that school is hard, living away from home is hard; you may be surprised to learn that other students are going through the same things. Talking about these issues can help to develop strong friendships that can last throughout college and even after.

If you don't feel comfortable talking with your friends about these issues or anything else, there are counselors on campus to help you deal with stress. As I said before, it is extremely important to determine what works for you and how to deal with your stress, and don't expect to have it figured out right away. It will take time.

Another thing to keep in mind is that the second semester is easier. You kind of know what to expect. After first semester, you know how college classes are different from high school classes and how to study. Also, you already know a lot of people so you don't have to go through making all new friends.

As for food science, I had a really good experience. I was an active member in the Food Science Club. I attended the meetings and went to other functions such as volunteering and socials. These were good opportunities to get to know people in food science. Attending socials helps to keep up your interest in food science. An additional benefit that I received from being a part of the Food Science Club was a summer internship. At a social event, I had the chance to meet someone from a company near my home town. She gave me her business card and told me to contact her, which I did. I eventually went in for an interview and tour of the plant. I worked for her for the summer following my freshman year.

An internship for a freshman is really rare. I was extremely lucky to have this opportunity. I was nervous going into it, however. I felt like I didn't have enough knowledge about food science in general and would be completely lost, but I figured I would do the internship anyway. I decided that what I wanted to get out of this internship was a better understanding of the food industry, an idea of what some different careers in food science are, and just to seek experience in general.

It was decided that my summer intern would be split between working in quality control and research and development. My first weeks I spent doing quality control in the plant. This was not what I expected and was my least favorite job. When working in the plant, I was really worried that I had picked the wrong career. I didn't like how hot and noisy that plant was, and the work was not what I thought I would be doing after graduating from UW. After talking to some people though, I realized this was just one aspect of food science and that there were other things I could do like research, food law, or even sales. The rest of my summer I spent working with the research and development team. I enjoyed this part of my internship the most; it was much more involved and related to what I wanted to do in the future. Starting off doing quality control was good for me because I got to see how a food processing plant works and how important quality really is. When the summer ended, I felt that interning at Masterson gave me a solid foundation on which I can build a future in food science.

Summary

Freshman year is the best time to get familiar with the university and to take advantage of all that is offered. It is the opportune time to use the available resources to figure out what you want to do and how to succeed in that field. Also, remember that you are on your own now and responsible for your own success, so be outgoing, learn where to find help, ask questions, and get to know people. In addition to studying, remember to take time for yourself and get involved with activities that you enjoy, so time management becomes an essential skill. Overall, it is important to make the most of your college experience, and it all begins freshman year.

Chapter 7
Is Food Science Right for Me? The Transfer Student

Leann Barden

Somewhere amongst classes, sporting events, extracurricular activities, seminars, and meetings with free pizza, college provides a time to discover your own interests. Virtually every student dreams of graduating with a degree that piques their interests and complements their career ambitions, but choosing the right academic major can be difficult, confusing, and time–consuming, but know that you are not alone. "Ninety percent of people don't know what they want to do coming into college, or they think they know what they want to do but find out later that it's not at all what they thought it'd involve. You just have to explore," said Luke Brosig, a senior majoring in food science at the University of Wisconsin-Madison (UW). This chapter of the book will guide you through the process and concerns other undergraduate students navigated when changing their academic major to food science. Although the interviewees in this chapter were all UW students, their stories transcend campus boundaries. When it comes to switching academic majors, only logistical paperwork differentiates one university from another. Students who decide to change their major experience the same exploration and decision-making processes, whether they attend a small technical college on the coast, a private university down south, or a research institution in America's Dairyland.

Discovering Food Science

Although TV shows like the Food Network's *Unwrapped* are increasing the public's general exposure to food scientists and their work, many undergraduates remain unaware of its existence, let alone what the major entails. So how did the undergraduates who switched to food science ever find out about

L. Barden
University of Wisconsin, Madison, WI, USA

R.W. Hartel, C.P. Klawitter (eds.), *Careers in Food Science*,
DOI: 10.1007/978-0-387-77391-9_7,

the major? Several students learned of food science simply by browsing online descriptions of all of the majors offered through their university. Many universities hold a Majors Fair, which provides an invaluable, commitment-free opportunity to learn about majors and their curricula and to question academic advisors and other students on hand. Departmental Web sites also offer information on curricula, course descriptions, departmental club activities, and faculty research. Search the site for names of academic advisors or a department chair you might contact with questions; if such information is not listed, ask the departmental office to refer you to an advisor. When registering for courses or perusing the departmental course descriptions online, check to see if an introductory/overview course is offered for food science. Student organizations provide another way to experience the culture of a department and to network with other students in the major. Katie Baures actively participated in the UW Food Science Club and Product Development team, becoming "fully immersed in the Food Science Department without being a major." Her participation in the club ultimately factored in to her decision to switch from engineering to food science.

What Will I Study and Learn in Food Science?

Food science applies chemistry, microbiology, and engineering to the world of food. "You should be prepared for a lot of different sciences," said Jen Baeten, a food chemistry graduate student at UW. "[Food science] is different from other majors where you focus on one type of science. But you especially have to like chemistry!" Most programs require food science students to take several semesters of food microbiology and food chemistry, as well as food processing, food engineering, statistics, nutrition, food law and regulations, and biochemistry. The prerequisites to these courses typically include general chemistry, organic chemistry, physics, biology/zoology, and calculus. Specialty electives in the major may include courses on packaging, confectionary science, dairy science, meat science, fermentation, sensory science, or flavor chemistry, although some universities may cover these topics as part of other courses. At UW, food science students choose between four possible degree options: International Agriculture and Natural Resources, Natural Science, tandem Biological Systems Engineering and Natural Science, or Agricultural Sciences—Production.

Will I Have to Stay in School Longer if I Switch Majors?

Whether or not switching to food science prevents you from graduating in 4 years depends largely on your previous major and respective college and on the timing of your decision. Students switching from another science major will undoubtedly have taken more applicable coursework than an English major, for example. Students switching from a major within the same college as food science will have taken applicable general education requirements, which can differ from college to college. Timing is also important. Switching majors as a junior or senior will probably require extra time to graduate in order to fit in all of your courses. Furthermore, some schools offer sequential courses (e.g., first and second semester food engineering) only one semester, so you may not be able to take many food science courses if you wait until the spring semester to switch. However, as a MasterCard commercial might say, "Tuition for an extra year of school: $6,000; finding the perfect major and career: priceless." Don't let the cost of an extra year of school force a decision you regret the rest of your life.

Andrea Roach, a former biology major, transferred into food science in her junior year, a move that required an extra year of school to graduate. "I love to cook and bake and became really interested in how food relates to your body," said Roach. Some of her food science classes introduced Roach to dietetics, and in her senior year, Roach decided to change her major to dietetics because it was "more focused in degree," even though the switch cost Roach yet another extra semester to graduate. Switching to food science "didn't bother me because I wasn't happy in my last major," said Roach. "I don't regret changing my major because food science opened a bigger window. I learned about more careers and saw different options. . . . I'm just really glad I did it because, if I were stuck in a major I didn't like, I'd probably be even more unhappy and end up going back to school in 10 years for a new major."

Dealing with Your Parents' Responses

While you ultimately need to find a major that challenges you and makes you happy, you may worry about your parents' responses, which can be quite varied. The parents of Jen Baeten, a former double major in horticulture and food science, supported her decision to pursue only food science. "My parents were particularly happy when I dropped horticulture because [horticulture] majors have problems finding jobs and don't get paid very well, whereas companies actually recruit food science students. And it's great that food science students don't need to pay for their internships. It's a good feeling to be wanted," said Baeten.

Tanya Zimmerman had trouble convincing her parents that food science was a better fit than biochemistry. "Food science was a tough sell to my parents, initially, because they didn't understand what it was. At that time, my dad was a medical technologist, and he thought food science was mostly microbiology and engineering. So he didn't see how I could go from biochemistry to microbiology and engineering." Once Zimmerman's parents better understood what food science involved, however, they encouraged and supported her pursuit of a degree in that major. "It's one of those degrees that not everyone connects what it is with the career outcome. . . . Of course, when I later decided to change my major to agricultural education, they were really blown away! (See "Personal stories" section.)"

Weighing the Pros and Cons

Making a list of the pros and cons of switching majors to food science may help some people to decide. While cons vary from person to person, consistently listed pros include high job placement rates, higher-than-average salaries for new graduates, undergraduate intern opportunities with major food companies, typically small class sizes, the opportunity to do undergraduate research (at UW and many other universities), and the wide range of career possibilities to suit every interest from food law and regulatory inspections to research to product development.

Morad Fadel, a former biochemistry major whose chemistry lab partner was a food science major, cited "the opportunity to pursue chemistry, other sciences, and business" as the most appealing aspect of food science. "There will always be a market for food scientists as long as there are people eating food," said Fadel.

Officially changing your major on paper can represent a big step, especially if you have any reservations about doing so. Realize, however, that the change need not be permanent; in theory, there is no limit on the number of times you are allowed to change your mind about majors! Changing your major and actually taking the coursework will give you a good indication of whether you made the right decision.

"It was a difficult decision at first," said Cathleen Radjenovich, who declared a major in food science second semester of her sophomore year. "Transferring caused me to stay an extra year because I switched in the middle of an academic year, and the classes follow a specific sequence. This was not a major factor in my decision to transfer though.

"I knew [food science] was the right major for me as I learned more and more about it. Taking an introduct[ory] food science class helped verify that this was what I wanted to do with the rest of my life because it gave me an overview of the opportunities I would have as a food scientist," said Radjenovich. "I am quite certain now that I made the right decision in my choice of majors. I thoroughly enjoy what I am doing now and find all of my classes extremely interesting. Internships and lab experience have also helped to verify that food science is the perfect fit for me. They allow you to work on so many different things that you would do once you are out in the workforce while having someone that will always be there for help."

Personal Stories About Transferring in to Food Science

Ultimately, every individual will have slightly different motivations for pursuing a career in food science, for finding their niche in the department. The following accounts illustrate the variables several UW students considered when officially declaring a major in food science.

Tanya Zimmerman, from Biochemistry to Food Science to Agricultural Education

"I was a biochemistry major but not thoroughly decided that's what I wanted to do. I liked biology and chemistry in high school and did well in both, so biochemistry seemed like a logical fit." At her college's orientation before the start of classes, Zimmerman toured some biochemistry labs with other freshman in the major. "I met this guy who had been studying the same protein for seven years, and he kept referring to this protein as 'she' even though a protein obviously doesn't have a gender. I thought, 'Please, God, don't ever let that be me!' After the tour, I went to get some ice cream at Babcock (the UW building that houses both the UW dairy store and food science department). Walking through Babcock to get ice cream, I saw the product development posters in the hall. After that, transferring to food science was a gut instinct," although Zimmerman did not officially transfer until the start of her sophomore year because she "wanted to be sure" she was making the right decision.

"Food science has a lot of science but with a really relevant application. I liked food science because people would be able to connect a lot more with what I do than if I were a biochemist just studying some protein. I went to my biochemistry advisor, who referred me to [an advisor in the food science department]. I didn't have to stay in school longer [after transferring] because biochemistry and food science paralleled really well."

Although she loved her coursework, Zimmerman eventually decided to change her major to agricultural education. "Transferring in [to food science] wasn't a difficult decision, but transferring out was because I'd never wanted to teach, but everything I'd done—through [my advisor], through the Food Science Club (on whose executive committee Zimmerman served as High School Outreach), through community volunteering—was all about education and teaching. Just because I liked it (food science) academically didn't mean it was something I wanted to pursue professionally for the rest of my life. I switched at the end of sophomore year—so I was already pretty far along with my classes—to [agricultural] education; from a program with good job placement, good advising, and competitive salaries to a program that offered none of those things."

Zimmerman, a fourth-year senior at the time of the interview, still wrestles with her decision. "Sometimes I wish I wouldn't have switched [from food science to agricultural education] so I would've had that background and professional experience. But ultimately, I would've gotten a degree in education. I just wish maybe I'd gotten my food science degree, worked in industry for about a year, and then gone for my teaching certification. I have a friend who did a similar thing with poultry science, and I see a lot of advantages to her qualifications now."

Luke Brosig, from Biology/Predental to Food Science

Before transferring to food science, Brosig was a biology major with the intention of going to dental school after graduation. "At the beginning of my third year, I realized I didn't want to be a dentist. I thought, '95% of people in [the College of Letters and Science (the largest college in the UW)] are biology majors. What am I going to do for a job?'

"I started searching through different majors on the UW Web site and read about food science. I emailed the contact person, and we met and talked more about the major and different opportunities available through food science; I liked product development and learning about the science behind cooking and food.

"I'd already done a lot of the [prerequisite] science courses as preparation for dental school, but switching [majors] still added on an extra year of school. I don't regret switching; it was a better long-term decision. And the extra year of school gave me the opportunity to do an internship, which led to a job. Now I'm starting my final year [of school], and I already have a job lined up for after graduation.

"Food science is so applicable to everyday life. There are so many times in class I'd think, 'Oh, that's why I have to cook to that temperature,' or 'that's why you don't need to refrigerate that.'"

Katie Baures, from Chemical Engineering to Food Engineering to Food Science

"Chemical engineering [is] more abstract, whereas working with food allows you to see results and relate to them in a common way. I don't mean that chemical engineering is abstract, but the microscopic level of chemical engineering was more difficult for me to be passionate about than food.

"I searched the College of Agricultural and Life Sciences Web site online and read about food engineering. I met with one of the food engineering professors, participated in the Food Science Club, and then changed majors in my sophomore year. I found that food engineering involved a lot of calculus, which I could do but wouldn't be content doing the rest of my life. I wanted more of a spread. As a food engineer, I was exposed to food science; I liked that food science incorporates engineering concepts but in a more rounded way. I knew I didn't want to sit at a desk job, and I like that food science gives you plenty of hands-on experience, whether you're in the plant or in the lab.

"My dad is an engineer, so he was supportive of my decision, but I think he always hoped I'd follow in his footsteps.

"To switch [majors] took at least a semester of talking to my parents, peers, and professors because I wanted to be sure [of my decision]. I'm a slow decision maker when it affects my long-term future. It will take me more than four years to graduate but only because I pursued other electives. In the long run, I made the right decision. I finally found a major that I could see myself doing later on in life."

Advice from Former Transfer Students

Sometimes the best advice comes from people who were once in your shoes, trying to decide whether or not to change majors. The following quotes are pieces of advice from students at UW who changed their major to food science.

If you are considering changing your major to food science, you should definitely speak with an advisor to find out what food science is all about. Advisors can help you decide whether food science is the major for you. Also, look around and do some research, even if it is just online. Speaking with an actual food scientist at a company is an even better idea. They can provide you with great insight into their career, and they are usually willing to speak with anyone that may be interested in the field.

—Cathleen Radjenovich

Take courses to keep your options open. For example, take two semesters of calculus, even if you only need one semester. A lot of people change majors and end up kicking themselves because they need to go back and take more calculus after they've forgotten everything from the first semester. Have a long-term vision beyond your current course requirements.

—Tanya Zimmerman

Talk to advisors and upperclassmen, and find out about job-placement, potential internships and the kind of course workload you'll have to take.

—Luke Brosig

Definitely get an advisor you like. Make sure you can talk to your advisor and feel comfortable doing so.

—Kelsey McCreedy

Learn more about different professions. Take into consideration the kind of classes you will be taking because you do want to be interested in what you will be doing for the next [several] years; you should like at least some of [the classes].

> Take courses that interest you. If you aren't exactly sure about everything, go with what you are sure about and see where it takes you. You can't have your whole life planned out when you're twenty.
>
> —Kelsey McCreedy

> Involve yourself—[go] to Food Science Club events, talk to professors, take introductory courses. People were so willing and happy to speak with me and help weigh my options. The more you find out about the major or the department in general, the easier the decision is.
>
> —Katie Baures

> Expose yourself to the major. Plant tours and internships are an ideal way to learn what food science is all about, but at the very least, talk to food companies, advisors and professors in the department, and other food science students. Having a food science advisor will probably be more helpful than having one in Cross-College Advising anyway, just because there are so many more undeclared's on campus than there are food science majors.
>
> —Tisha Yancey

Moving Forward

The stories in this chapter were meant to guide all undergraduate students—not just those attending UW. Armed with advice and reflections from students who were once in your shoes, weighing the pros and cons of abandoning one academic major to pursue another in food science (or double-majoring), you must now make your own decision and move forward to embrace the inevitable changes. Learn from other people's mistakes and successes by finding the commonalities between their situation and your own. And remember, "Tuition for an extra year of school: $6,000; finding the perfect major and career: priceless."

Chapter 8
Landing an Internship

Leslie Selcke

The Decision to Pursue an Internship

The decision to pursue an internship can be just as difficult as finding one. Choosing between completing on-campus research for a professor and gaining industry experience through an internship can be a struggle. Most advisors, professors, and professionals alike will agree that an internship is a must, and employers consider internship experience as one of the most important factors in hiring college graduates. The knowledge and skills attained through lectures and laboratory courses only lay the foundation for understanding the food industry. Actual work experience is necessary to gain a true sense of what the industry is all about. Students can start looking for internships as early as their freshman year of college. There are many opportunities available, it is just a matter of finding one that suits your interests and then pursuing it.

It is important to note that there are several different types of work experiences available. Internships are often termed summer work experiences, but there are several differences between the two. Internships are generally more formalized, involve determining objectives, and require writing a final report or giving a final presentation reflecting on the experience. A summer job is more or less working for pay. It may be helpful in learning more about the food industry, but formal objectives are not always apparent. Some companies have more formalized intern programs than others. The formal ones tend to have social and professional programs, while at other companies you could be working more on your own. For example, larger corporations tend to have social programs such as going to a baseball game

L. Selcke
University of Illinois, Champaign, IL, USA

R.W. Hartel, C.P. Klawitter (eds.), *Careers in Food Science*,
DOI: 10.1007/978-0-387-77391-9_8, © Springer Science+Business Media, LLC 2008

planned for the interns. The programs promote social interaction between the interns and help develop stronger relationships. A professional program may consist of organized lunches where interns learn about the company's history or even how the patenting process works from current employees. However, at smaller companies, there may be fewer interns and resources available to plan programs such as these.

An internship provides an avenue to see what a full-time position in the food industry will be like. It gives insight to the type of position the student desires to have upon graduation. It also teaches the student how to apply classroom experiences to their future careers. Internships can help build confidence as well as allow students to mature professionally. Listed as follows are some of the main benefits of completing an internship.

Key Benefits

- Learn about the food industry from the inside
- Determine if this career path is the right choice
- Improve business savvy and etiquette
- Boost self-confidence in personal abilities and skills
- Further develop resume with food industry experience
- Build professional contacts for future job search
- Apply academic learning in a practical, hands-on way
- Potentially obtain a full-time job offer
- Enhance public speaking skills
- Collaborate with cross-functional partners to learn about the different aspects of the company
- Receive advice about pursuing further education either before or after entering the work force
- Develop new technical skills and build credibility
- Meet students from other universities around the country and the world

Before starting, make sure to have a clear understanding of the reasons to obtain an internship. The motives for obtaining an internship vary from person to person. For example, one individual may be looking to be an effective communicator while another student may be focused on improving his or her technical skills. Either way, by having a clear vision about what there is to gain students can focus on areas of improvement and start working on them from day one. It would be a shame for an intern to realize that at the end of the internship, they did not challenge themselves to improve social and technical skills. Students are much more likely to have an enjoyable and successful internship by establishing goals from the beginning.

How to Find One

There are many routes to finding an internship that suits your needs and interests. The two main resources available are through universities and the World Wide Web.

Opportunities for internships can start to appear on campus as early as September. At large universities, many companies plan informational meetings through the food science department or Club. At these info nights, industry professionals talk about their company's history, the internship experience, and even the outlook of the company. Some have previous interns speak and discuss their personal experiences. The company may collect resumes at the meeting and host interviews later or ask students to submit their resumes online. Either way, this is a great opportunity to make contacts with companies that have a great relationship with the university.

For an informational session, dress in business casual attire and act professionally. Avoid talking to people sitting nearby during the presentation, but feel free to ask questions to the representative about the company and their internship program. Possible topics that may need further clarification could be the size of the program, the number of intern positions available, if the internship is paid or volunteer, the locations available to work at, the types of positions they have to offer, and when the company will be conducting interviews. If previous interns speak about their experiences, feel free to approach them afterward with more questions as well. Also, most former interns are willing to share their email address for questions that may arise later. If time is allotted, it would be helpful to speak directly with the company representative after the presentation is complete. After turning in a resume, ask for their business card or contact information. It is also wise to follow up after the informational session by sending a thank you email to the representative(s) one has spoken with.

Another avenue for finding an internship is to attend career expos held through different colleges on campus. If looking specifically for a food science internship, both the engineering and agricultural colleges provide the most opportunities. Make sure to dress professionally and bring plenty of resumes. Before the event, it is best to look at a list of the companies attending the expo and plan which booths are worth visiting. Then, do some basic research on the company's history, size, and location. Having a base knowledge of the company demonstrates a strong interest in the corporation and provides great conversation starters. When talking with companies, be sure to make a great impression. Dropping off a resume without spending the time to get to know the representatives will not make a strong impact. Students should spend time talking to the company representative rather than focusing on the next booth to visit. For a student to make their presence known, he or she should display enthusiasm and genuine desire for wanting an internship. It may be helpful for a student to prepare a short introduction that discusses the unique qualities that he or she has that could benefit the company. During career fairs, companies speak with hundreds of students looking for employment. It is critical to stand out for further consideration. Make sure to send a note thanking the company representative for their time.

Other opportunities on campus include the advising department and online resume uploading. Keep an eye out for flyers and emails containing internship opportunities. There is generally a higher chance to obtain an internship if the company is going to the university for students. This indicates there are specific positions that need to be filled. Also, some schools provide a Web site where each student can make a profile. Students can upload a resume and allow companies to search for it. In the profile, students can narrow down opportunities based on specific majors, positions of interest, and locations. Once an opening is available, the company can either contact the student directly or set up an interview slot.

An additional source for internships is to network. In any field, networking is key to learning about job openings that may not be publicized. Talk to as many people as possible that could help in the search for an internship. Also, describe to them the type of internship being sought out. Department faculty, friends, alumni, parents, relatives, and family friends are great to talk to. Surprisingly, most people have some sort of contact with professionals in the food industry. If meeting with someone who works directly for the food industry, try to set up an appointment with them at their office. Tell the person how their name was obtained, and that the goal is to not necessarily obtain an internship with their company, but to gain additional information about career options. Prepare a list of questions ahead to time and always send a thank you note afterward. Listed below are some possible questions to ask:

- Why did you choose this career path?
- What do you do during a typical day?
- What types of experiences do you recommend before searching for a full-time position?
- What qualities and skills are essential to being successful in this field?
- What are the available entry-level positions for college graduates?
- Do you recommend any classes I should take to further develop my professional skills?
- Are there any aspects of my resume that are particularly strong? Weak?
- Do you know any other professionals in the industry who would be willing to talk with me? May I mention your name when contacting them?

An alternative path for finding an internship is on the Internet. There are hundreds of career exploration opportunities online, and many include searches for internships. Web sites such as *CareerBuilder.com* and *MonsterTrak.com* are databases oriented to college students. However, it is important to realize that everyone around the country has access to these sites and companies may receive hundreds of applications for only one position. Therefore, a better route would be to visit the Web sites of major food companies directly. Most have a list of job/internship openings that are listed by location and type. Resumes can be uploaded or emailed directly to the company's Human Resource Department. If impressed by a submitted resume, a company may look for openings that fit the applicant's wants. However, if a student already has contacts at this company, they should use them to get their resume to the right person.

The Institute of Food Technologists (IFT) can also provide a way to find an internship. Although there is a membership fee it is a great organization to make contacts around the country. The directory can be used to look up professionals at certain companies or alumna from universities around the globe. There are individual sections divided by location that students can join for a small fee. The sections provide plenty of opportunities for employment. Attending section meetings to make contacts and network is also important. There are also a variety of divisions within IFT for special interests such as Dairy Food and Citrus Products. These divisions are available to members as well as student members. In addition to the divisions, the Institute of Food Technologists Student Association (IFTSA) may provide other means of finding an internship. Check their Web site at www.iftsa.org for more opportunities.

How to Evaluate Different Offers/Opportunities

Evaluating more than one offer for an internship can be intimidating. While money is always a big motivating factor, it is imperative to look deeper than just the base pay to find out all of the benefits. Here are some factors to consider:

Size: The first aspect to critique is the size of the company and their internship program. Some companies will have hundreds of interns, while others will only have a few. Living alone or being surrounded by dozens of interns can change the way to look at an offer. It can also indicate if an intern will be working in one division or several. The size of a company and intern program may also indicate how much individual attention the intern will receive. If worried about this, ask the company about the supervision that interns receive throughout the internship. For example, large companies may pair interns one on one with a direct manager. However, small companies may have one individual in charge of a small group of interns.

Supervision: Depending on the size of the company, various levels of supervision may be given to the intern. Some companies give interns a direct manager to report to while others have interns work on their own. With a direct manager, interns generally meet weekly to discuss the progress of the project, review formulations, go over any questions that may arise, and allow the manager to offer advice. This type of supervision is often more involved in the project and is constantly up-to-date on the progress. Other programs may have their interns meet with a supervisor only in the beginning, middle, and end of the internship. The intern is expected to make decisions on his or her own and know when to seek out additional help from other experts in the company.

Project Load: Be sure to ask what type of work will be completed during the internship. Smaller companies may have their interns working on a variety of projects in different departments while larger ones may assign one project to be taken from start to finish.

Position: Along the same lines as the project load, make sure to fully understand what type of position is being offered. A product development position is much different from having a quality assurance internship. Similarly, working in sensory would entail a completely different role than a research position.

Company Location: Check to see if the company has more than one location to work at. This could provide opportunities for relocation in the future. Also, moving out of state for the summer may not work for some students but could appeal to others.

Working Style: Make sure to understand from the beginning if the internship responsibilities include working with others or independently. A person oriented to working in groups would struggle with a position focused on independent assignments.

Living Accommodations: If unable to live at home or at school during the internship, ask if the company offers housing for the duration of the internship. Some subsidize the cost while others provide a stipend. Also, check to see if the company will pay for relocation in the form of mileage, airfare, or other traveling fees if moving long distances.

Hours: Find out ahead of time what type of hours the work day consists of. Some companies are flexible with the start and end time of the day, but others have more strict business hours. Most companies will expect interns to work a 40 hour week.

Summer vs. Semester: A summer internship is often easy to fit into any college schedule, but a semester program allows for more time in the workforce. As long as classes can still be completed on time, a full semester internship will give students better insight on how a food product is developed from start to finish. A semester is a realistic timeline for a project to be completed in the real world.

Duration: Ask about the length of the internship from the beginning. Some will allow students to work 10–12 weeks during summer vacation, but others require more time. Keep in mind that studying abroad for a few weeks in the summer may not allow enough time for a student to complete an internship.

Financial Aspects: There are a variety of financial aspects to consider when evaluating internship offers. Some internships are nonpaying, but provide students with amazing learning experiences. Other companies may pay on an hourly wage or put their interns on a salary.

Future Employment: Often students will find that an internship may lead to a future job offering. If a student is struggling between two offers, it would be wise to explore and evaluate the positives and negatives of each company, considering what would best suit their personality, needs, and future goals.

Take the time to look through each offer carefully. Most companies will allow from several weeks to over a month for a response to their offer. As mentioned earlier, look past the base pay and make sure that the internship will help develop personal and technical skills to prepare for a full-time position. If anything remains unclear about the offer, be sure to contact the companies' Human Resource Department to find the answers. If unable to answer any questions, the representative will put you in contact with a person who can.

How to Connect Internship Learning to Academic Learning

An internship helps connect academic learning to industry experience. The two go hand-in-hand and complement each other for the benefit of all students. Academic learning lays the foundation for any food-related internship. In return, internships reinforce classroom knowledge by actual work. Not everyone will use their food science learning as an intern, but others will find extensive use for it. The amount utilized will vary based on the position and the project assigned. Also, while good grades are important in college, they are not absolutely critical to obtaining an internship. Not all companies focus solely on good grades but look for a well-rounded individual, one who participates in more than book learning.

At the start of an internship, all interns are expected to have full knowledge of the basic principles of food science. Even if you are a freshman or sophomore, make sure to have a broad understanding of what food science encompasses. It would be helpful to read journal articles pertaining to the area you will be working in. Also, students can contact their manager before the start of an internship and ask for supplemental reading to help with the project he or she will be working on. It can be helpful to keep class notes at hand if any specific questions come up. Most positions will teach interns everything they need to know, but having a base knowledge is crucial to seeing the big picture. For example, a lot of equipment used in laboratory classes will be utilized in an internship. The hunter colorimeter and moisture analyzer are two very common instruments used in a food science industry. Also, a sensory intern will know how to conduct evaluation sessions, use proper terminology, and summarize results statistically based on information from the classroom.

It is also wise to keep a journal of weekly activities and learnings. A journal keeps accurate records of what has been completed during each week of the internship. It will keep you up-to-date on the progress of your internship. Also, some schools require students to keep a weekly journal to receive class credit for their internship experience. Others allow a paper or presentation summarizing the experience for credit. If interested in learning more about receiving class credit, speak to an academic advisor or the food science department.

Besides applying previous classroom learning to an internship, knowledge gained from an internship can provide opportunities for better learning in later classes. For example, an intern who worked with a spray dryer may learn the mechanisms of its operation in a processing class the following semester. Also, product development internships will teach students the proper timeline and procedures for creating a new product, which students may apply in their senior year capstone course. Working with previous formulas in the pilot plant will also provide technical knowledge when learning how to use the formulas in a future food engineering course.

While team projects are a large part of both academic and internship experiences, in the industry they can be slightly different from the classroom. When working on a team project for a food company, each person has specific roles assigned to them. Usually a team project will consist of individuals from several positions such as a food scientist, a sensory scientist, a marketer, and a technical expert. Each person is expected to contribute their portion of the project on time. Unlike school where the hard workers have to complete the work of the less involved team member, repeated failures of being a team player in the work force can cost an individual their job.

Internships are a great way for students to apply their knowledge gained in the classroom. They can supplement topics already studied, or provide insight on classes to be taken in the future. Either way, an internship will greatly strengthen a student's familiarity in many areas of food science.

Tips for Making the Most Out of It

There are countless tips for making the most out of any internship. To start off, research the company before starting the internship. It is helpful to understand the structure of the company, their products, and the size at the start of the internship. This simple information will help you see exactly what your role is in the company. Listed below are tips for succeeding at any internship:

Set Goals: Make a list of realistic goals for the course of the internship. Focus on personal development, technical skills, and other aspects that need improvement. It would be helpful to ask for a second opinion from a supervisor and/or campus advisor on the goals established. Both individuals can provide useful feedback. Also, periodically check back at the list and update what has been accomplished.

Keep a Daily To-Do List: It is imperative to keep a to-do list on a desk calendar or notepad. Cross off what has been completed and stay on top of what needs to be accomplished. Something as simple as forgetting to submit a prototype for micro testing can set a project back a couple of weeks or delay an important meeting.

Ask Questions: Take advantage of being surrounded by experts in the food science field. No intern is expected to know how to do everything, so ask questions from day one. The more questions asked, the easier it is to start working on a project. Also, asking questions can prevent you from making a costly mistake. For example, using the wrong formula for a product will waste valuable resources or operating a piece of machinery in the pilot plant incorrectly can cost thousands of dollars in repairs.

Write a Weekly Update: Every Friday, write a weekly update to reflect on the work that has been completed that week. Not only will this help you stay on track, but it will allow you to quickly compile your work in chronological order for a final presentation or report.

Meet with Supervisors Regularly: At the start of the internship, interns should schedule weekly or biweekly meetings with their direct supervisor. The meetings can be rescheduled if something comes up, but it is important to get on their calendar right away. The topics discussed at the meeting may include progress reports, expectations for the following week, questions/answers on the project, and accomplishments of the internship so far. Listen carefully during the meetings because some of the best advice will be given.

Get to Know the Other Interns: Make sure to spend the time to get to know the other interns at the company. Some may have worked at the company before and know the ins and outs of succeeding. Others may act as a source of information about the company and departments within.

Participate in Activities Outside of Work: Take advantage of all the opportunities a company may have to offer. Some companies have intramurals sports leagues, while others may give away tickets to nearby sporting events. Go to the special seminars during lunch breaks and learn about new, emerging technology in the field. Getting to know other employees and interns outside of the workplace can be extremely rewarding.

Laboratory Notebook: If working in product development or research, keep an accurate and up-to-date laboratory notebook. Having to go back at the end of the internship and add every experiment completed is extremely stressful. Spending 15–20 minutes a day will be helpful and beneficial when working on a final presentation or report. It is also advantageous to make electronic files of formulas, outlines, instructions, etc., because it will be much easier to transition the work to the next person assigned to the project.

Network: Schedule "meet and greets" with as many people as time permits. Meet and greets are informal meetings with an intern and an employee at the company to learn about their job. To schedule one, you should either call or email the professional and schedule a 30 minute meeting with them. You should explain that you would like to learn more about the professional's position and experiences at the company. Make sure to schedule the appointments early in the internship. Also, if unsure with whom to make meetings with, ask a direct manager or supervisor. These experiences can provide a great deal of knowledge. Meeting experts in a specific area can be helpful later on when questions about a specific ingredient or process arise. Also, it is a great way to learn about the different positions that are offered within the company. Building strong, professional relationships with managers and employees in the company is imperative when being considered for future employment.

Keep in Touch: After the internship is completed, take the time to keep in touch with close contacts made during the internship. A quick email once a month is enough to keep those close ties. This may be crucial to getting a job in the future. It may also be beneficial if a question about a school project comes up and industry advice is needed.

Resume Update: At the end of the summer, you should bring in a copy of your resume to review with your manager. This is important so that no proprietary information is released. It is also helpful to get advice on how to add all the contributions, skills, and knowledge attained onto the resume.

By following these simple pieces of advice, you will come out ahead of the group. Keeping track of project work, due dates, and meetings will prevent high levels of stress from accumulating. An internship can be an extremely educating experience, but one must put in the time and effort to make the most out of it.

Internships as Interviews for Full-Time Employment

An internship can easily lead to a full-time position in the food industry. Working for a company allows them to see the strengths and weaknesses of their interns, as well as public speaking skills, work ethic, and professional behavior. After investing a lot of time and money into interns, a company will no doubt want to have their interns come back if they do great work. All presentations and meetings can act as preliminary interviews. If interns can act themselves and gain respect from the people around them, current employees will only have positive things to say when evaluating interns for full-time employment.

It is important to realize that how you treat everyone around you will say a lot about who you are. Strong communication skills, work ethic, time management, and organizational skills are all factors that employees notice. If you can succeed in all of these areas, the company will be easily impressed. It is also important to realize that in order to get a job, the project being worked on does not have to be a complete success. As long as you follow the proper steps and work beyond the minimum, you will have a chance at a full-time position. Many projects worked on never make it to the grocery store shelf. As long as you complete your work and keep good records in a lab notebook of the successes and failures, you will have done a great job.

The process of hiring interns for full-time positions varies from company to company. Some meet after the interns have left and discuss their work during the internship, evaluate their contributions, and review their resume. Other companies have an exit interview for interns with a Human Resource representative to determine if they will be invited back. Whatever the process may be, just make sure to leave a great impression at the end of the internship and hopefully a job will open up for after graduation.

Conclusion

Over the past few years I have grown immensely as a food scientist. Completing two summer internships provided me with the opportunity to look at the food industry from the inside. I was able to complete bench top work, gain experience in the pilot plant, collaborate with cross-functional partners, visit a product launch, and meet many intelligent individuals with a passion for food science. Not only did I learn how to act in a professional manner by gaining confidence in my communication skills, but I also enhanced my technical skills and ability to "learn on the fly." The internships provided me with the ability to visualize situations discussed in the classroom as a larger picture in the food industry. By providing goals for myself and staying organized by keeping accurate records and organized files throughout the summer, I was able to make the most out of my internship and not waste any time. With strong organizational skills, I was never stressed or quickly putting something together last minute. Also, by meeting with many employees from the company, I was able to get a good feel for what the company stands for, the atmosphere, and the type of individuals the company hires. Through hard work, dedication, and a drive to succeed, I was able to show the company the strong skills I have to offer. Fortunately I received a job offer and look forward to working for them full-time upon graduation.

While the search for the perfect internship can be intimidating realize that internships provide valuable experiences. To find an internship, make sure to start the search early. Participate in as many opportunities on campus as possible, which put you in contact with other food science students and professionals in the industry. Also, take classes seriously and understand that they lay the foundation of your knowledge as a food scientist. When evaluating internship offers, realize that money is not the only motivating factor. Location, project, and size can have an impact on the experience. While at an internship, make sure to set goals, stay organized, and act professionally. Internships themselves can act as an interview for full-time employment. Also, network with as many people as possible—from the internship, university, friends, and family—for help with the future job search. Finally, be yourself and aim to find your fit in the food industry. Not all internships will be exactly what you see yourself doing, but talking to others from the field can help you find something that fits your needs.

Chapter 9
The Leadership Case: Student Perspective on the Value of Leadership Skills

Tanya Zimmerman

Defining Leadership

Contrary to the simple heading, defining leadership is a complex task. It is easy for us to recognize leaders, but prescribing the skill set needed to be a leader can be more challenging. The difficulty in defining leadership is inherent in the nature of the role itself. We perceive leadership to be an elite skill, yet anyone with enough motivation can become a leader. Leadership is a learning process that is ever developing. It is not a singular event, nor is there a sole list of skills that one must possess to acquire the title of 'leader.' There are numerous leaders in our world, each working toward different goals with different people. The challenge in both defining and realizing leadership comes in recognizing the many leadership opportunities that already exist in our day-to-day lives, and being willing to constantly grow, acquire new skills, and adapt our behaviors to appropriately match each new leadership situation we encounter.

Despite the diversity that leadership embodies, there are skills on which all leaders rely. The following list captures some of the most valuable skills and ideas for students in leadership roles:

- Invest in building relationships. By taking the time to understand yourself and others, you value everyone's contributions more.
- Approach tasks by realizing possibilities, setting goals, outlining plans of action, and recognizing the larger purpose of your work. Be adaptable and prepared for change; it's often an inevitable part of leadership.
- Strengthen your communication skills, both in expressing ideas and listening to others. There are a variety of ways to communicate ideas.

T. Zimmerman
University of Wisconsin, Madison, WI, USA

R.W. Hartel, C.P. Klawitter (eds.), *Careers in Food Science*,
DOI: 10.1007/978-0-387-77391-9_9, © Springer Science+Business Media, LLC 2008

Work to include communication techniques that best meet the needs of your group.

- Enable others to take on a meaningful role in the process. This not only creates valuable stakeholders and raises accountability, but it also helps prepare 'the next generation' of leaders.
- Remember that leadership is a process, not a singular event. Find learning opportunities in all situations, even those that aren't the most rosy. By taking time to reflect, you can build your own leadership skills while preventing adverse situations from repeating themselves.

Why Invest in Leadership?

If we reflect on those who have had a positive impact on our lives, it is likely they embody many of the skills listed earlier. When considering the value of leadership and its importance, take time to think about how your life has been improved by the leadership of others. In many ways, leadership skills are more aptly categorized as interpersonal skills. By taking time to engage in leadership learning, you'll strengthen these skills and likely enhance your relationships with others.

As you move through your collegiate years, you will also find that leadership skills are a major asset to your academic and career success. From group projects to internships and extracurriculars, you'll find that you often need to interact with others and work as a group to accomplish goals. Food science careers emphasize team-based work, but you'll find that strong leadership skills are useful in all career paths. Such skills also help bolster your decision-making abilities. You become better equipped to evaluate possibilities, resources, and teammates; communicate ideas; and manage the process. At the end of the day, a group's chance of success is impacted by the leadership skills of its members.

Employers also value leaders, and seek out employees who have invested time and energy in developing leadership skills. Talk with professionals in your field, alumni, and professors to determine what specific skills employers want. In addition to seeking out opportunities to become proficient in these areas, consider how you can communicate these skills during a job interview or a career fair. Leadership development is a valuable aspect of career planning. Students who invest in acquiring these skills often open the door to an array of exciting professional opportunities.

Ultimately, developing leadership skills improves the wide array of communities that people belong to, from their town, church, organization, or workplace to their dorm, family, or even peer group. Leadership skills allow you to create vision and goals for community improvement, strengthen relationships among members, and also be a positive agent of growth and change. In essence, developing and utilizing your leadership skills allows you to pay forward the positive impact other leaders had on you.

How Do You Acquire Leadership Skills?

Similar to the diversity of leaders, acquisition of leadership skills is like a mosaic. There is no single class, event, or effort that will make you a leader. Instead, you must find learning opportunities in varied places, piecing together activities, lessons, and skills to develop yourself as a leader. College abounds with opportunities to "stretch" yourself in many ways, including as a leader. Utilizing the ideas from Chapter 4, along with the ideas presented here, provides a great base for developing leadership skills.

As a college student, one of the easiest ways to get involved is to participate in an extracurricular activity. Most schools have a variety of student organizations related to different academic programs, as well as a multitude of clubs for other student interests. Try out several groups to find one or two that fit your needs and interests. Recognize the variety of contributions you can make to an organization, and the time and other commitment necessary to be a member. Once you find a group that is a good fit, get involved! Most clubs are always looking for new members to help with events or serve on committees, so you are likely to have plenty of hands-on opportunities to develop leadership skills. Volunteering can also be a great chance to develop leadership skills, while giving back to your community. Some campuses have an office that coordinates volunteer opportunities. If your campus doesn't, consider contacting your school's student services office; they may be able to help you find a great volunteer placement related to your interests.

Many universities also have leadership education programs that are specifically tailored to help students develop the skills needed to succeed in leadership roles. These programs may range from a one-day workshop to a long-term training course. However, all can add substantial value to your education, specifically challenging you to think about leadership skills and equipping you with the knowledge and tools to be a better leader. The key to the success in these programs, however, is making use of the information. Apply what you've learned to your own leadership roles or involvement opportunities. Take time to evaluate your own strengths and weaknesses and make a plan for improvement based on what you've learned. This could include taking on a leadership position in a student organization that would challenge you to grow; taking part in other programs, such as the communications skills group that Annie joined in Chapter 4; or even reading and reflecting on different leadership theories.

College is also a great time to learn from the wealth of people around you. Seek out chances to build relationships with other leaders on campus, such as professors, advisors, upperclassmen, alumni, or leaders in your profession. By talking with them about their experiences and advice for students, you are able to get real-world feedback on the importance and impact of leadership.

Often, these relationships can also foster leadership learning opportunities, ranging from a new role in a student organization to a summer internship. Subsequently, forming these connections is an example of the overlap between leadership and professional development.

Beyond the array of leadership opportunities you can find in extracurricular activities, don't forget about the learning potential in your academic programs. Think about the group projects or lab teams you participate in and evaluate leadership strategies that improve the group's functionality. Consider using general education courses to strengthen your skills in areas like communication or ethics, both of which are key aspects of leadership development. Also take note of freshman seminars or related courses at your university, which are specifically designed to help students transition to campus life and present opportunities for making the most out of college. While each university is unique in its programs and curriculum, utilizing resources like your academic advisor, student services office, and leadership education programs can help you find ways to develop your leadership skills even within the classroom.

Most importantly, as you develop as both a person and a leader, you often need to step outside of your comfort zone to truly grow. Like the students in Chapter 4, challenge yourself to pursue new opportunities. As a student, if you love to cook, but there is no cooking club on your campus, take the steps to form one. By taking action, you are able to realize your skills and potential in ways that are tough to capture from reading a book or attending a workshop. As you pursue these new opportunities, set goals that challenge you to improve yourself, and utilize the people and resources around you to reach them. As you move forward, reflect on whether you are still moving toward your goals, and what this development means in terms of your leadership skills. It can be helpful to seek feedback from others regarding your performance, especially when you are in leadership roles. Although it can be difficult to solicit and receive feedback, it is a key component of self-improvement, allowing you to grow in ways you may not see independently. Ultimately, if you are motivated and passionate to become a better leader, you will find opportunities to develop your leadership skills in all aspects of your life.

Putting a Face to These Ideas

Writing this chapter on student leadership is an opportunity that I couldn't have imagined in my freshman year of college, or even at the beginning of my senior year. As a student, I've transitioned through different academic programs, changed career goals, and seen different friendships bloom and fade. In short, a lot has changed. Nonetheless, in reflecting on my time as an undergraduate, I see that each relationship and decision has shaped where I am going today, especially with regard to my perspectives on leadership.

I came to a campus of over 40,000 students after graduating from a moderate-sized high school where I knew everyone. Needless to say, being one in a sea of freshman, all swimming aimlessly toward graduation, was slightly overwhelming. I didn't actually do much in my freshman year, and instead tried to just learn my way around campus. I realize now that this period of acclimation was an important part of my college transition; however, it was not until I took a chance and got involved that I really felt connected to campus.

My first true leadership experience happened during my sophomore year, when I volunteered to serve on a committee planning a retreat. The advisor for the committee gave all of the students some substantial planning roles, which ultimately would impact the success or failure of the retreat. Being given this amount of responsibility, especially as an unknown underclassman, which boosted my confidence. It was that experience than made me feel like I could take ownership in the leadership process on campus, and subsequently make a difference. Although I went on to organize bigger events and projects over the next few years, nothing impacted me as much as that planning committee. It wasn't arranging for the caterer that was meaningful; it was the work of my advisor, who enabled me to take responsibility, making me feel like my contributions were vital for the group's success.

That event was the tipping point for my student involvement. I became very involved in the student organizations in my college, both, those related to my major, as well as those related to leadership. Over time, I served in these organizations as a representative, committee member, and an officer. I saw what leadership strategies worked well for my peers, as well as strategies that weren't so effective. As I moved into leadership positions during my junior and senior years, I tried to model the kind of leadership that inspired me. Often, this meant trusting members I didn't know very well to carry out responsibilities for the organization. It was nerve-wracking at first, as guiding someone new can be more time consuming than doing it yourself. However, the long-term benefits were well worth any related costs. Enabling students to take on these leadership roles generally inspired them to be active members who I could trust and depend on in the future.

Of all the activities I participated in as an undergraduate, the most influential was a leadership certificate program offered by my college (www.cals.wisc.edu/students/leadership/Leadership˙Certificate.php). It was a long-term program that emphasized self-reflection and understanding of leadership, based on conceptual and behavioral competencies. As a student, I saw the value of such a program, but I did not appreciate the impact that it would have on participants until I began preparing my own evidence for the leadership competencies. The program challenged me to see how all aspects of my college career contributed to my growth and progress as a developing leader. It pushed me out of my comfort zone, challenging me to communicate my strengths and weaknesses, and develop my leadership voice. I left feeling like I could draw out core leadership ideas and values from all of my experiences on campus, both the positive and the negative. It brought to light new value in situations I dismissed in the past, and caused me to place more stakes in finding learning opportunities in the new experiences I encountered. In many ways, although the program did not count for degree credit, it was the apex of my collegiate learning. Although not all students are fortunate enough to find such valuable programs at their own schools, I encourage you to seek out opportunities to self-reflect on the meaning of your experiences and develop your own leadership philosophy. The resources to do so are boundless, but you'll never recognize the value they have, unless you take the time and energy to discover it.

To Sum It Up

Leadership is a long-term process, but one that is worth pursuing. By acquiring the interpersonal, communication, and organization skills central to leadership, you equip yourself with the tools necessary to impact your profession, community, and relationships with others. Developing leadership skills is not easy, but with the enthusiasm and commitment to do so, everyone has the capacity to lead. Utilize the resources on your campus and in your community, and challenge yourself to recognize the learning opportunities that already surround you. As you move forward in your leadership development, reflect on your progress and goals, always continuing to strive to become more like the leader(s) who inspire you.

Part III
The Graduate Student Experience

Chapter 10
Finding and Getting into the Right Graduate Program

Laura Folts

Making the Decision

Students choose to pursue a graduate degree in food science for many reasons. These may include: a means of entry into food science, future job requirements, or expansion of food science knowledge. Regardless of reason, the chosen graduate program should provide the student with the skills necessary to meet his or her goals. If the goal is to obtain a specific career, the program should be selected and tailored to fit the needs of that job. Or, the student may wish to develop a greater understanding of a particular area of food science, and thus develop an expertise. Additionally, many students will choose to pursue a graduate degree in food science after obtaining a completely different undergraduate degree and the graduate program will include the fundamentals of the science.

While not everyone chooses to go to graduate school for the same reasons, it is important that once the decision to go is made, careful thought is given to both the school and the research focus. Graduate school involves intense research in a given area of food science—it is important that the focus, as well as the school, is chosen with care to provide for the best experience possible.

Identify Graduate Programs

There are many resources available for identifying graduate programs. For those currently enrolled in an undergraduate program, a good place to start is to talk to an undergraduate advisor. He/she has experience with graduate

---

L. Folts
University of Minnesota, St. Paul, MN, USA

R.W. Hartel, C.P. Klawitter (eds.), *Careers in Food Science*,
DOI: 10.1007/978-0-387-77391-9_10, © Springer Science+Business Media, LLC 2008

school on both a personal and professional level and may have suggestions about what programs to consider. More specifically, undergraduate advisors may know faculty members at other schools and may be able to make recommendations on the applicant's behalf. Additionally, the advisor may be able to recognize strengths and weaknesses in both the student and the graduate programs and direct the student to a better fit. While the undergraduate advisor is a great resource, other resources are also available.

Another way to learn about graduate programs is to talk to people in industry. Many food science students do internships, which provide excellent contacts. Contact current or past coworkers about what schools they considered and which one they chose. Since not everyone is looking for the same graduate experience, it is a good idea to utilize the experiences of more than one person. Word of mouth is a great way to get the inside scoop on what people think about specific advisors. Some suggested issues to address are ease of working for the advisor, ability to get questions answered in a timely manner, what do students think of that advisor, and how are they regarded in the industry. Also use industry resources to get information on the department, school, and city. Coworkers may be able to provide personal experiences and preferences about a school/city that would not be available in a campus brochure.

The Internet is also a powerful resource for exploring potential schools. Two suggested tools are individual school Web sites and the Institute of Food Technologists (IFT) graduate program directory (http://www.ift.org/cms/?pid=1000624). The IFT site provides a quick summary of schools that offer graduate degrees in food science. The summary includes varied information on the program and a list of the faculty and their research focus. Individual school Web sites can provide more specific information about the university as a whole, the department, and faculty.

Graduate schools have more to offer than just the academic curriculum. It may also be useful to consider the city, state, and campus of a graduate school. For most of these factors, personal preference is the most important consideration; however, the cost of living may be of significance. Schools in larger cities may be more expensive—maybe not in tuition, but in housing and food costs.

Finally, not all graduate programs are considered equal by future employers. Some companies may have selected a few schools from which they regularly recruit. This does not mean that they may never hire anyone from a school not on their preferred list, but it might make getting the job a little more difficult and is something to consider when choosing a program. Information about affiliations between schools and companies can sometimes be found on the company's career Web site. Or, use industry contacts to gain this information.

Initial Contact

Once the graduate programs/schools of interest have been identified, it is a good idea to begin contacting advisors to discuss the possibility of working for them. This process saves time and money in the application process since the prospective student only applies to schools in which a faculty member is interested in working with them. Additionally, it may be easier to get accepted if a faculty member is aware of the student and would like to work with them.

It is generally a good idea to initiate contact with a potential advisor via email. Faculty members are busy and some may be caught off-guard by an unexpected phone call. However, because they are busy, a follow up phone call to the initial email may be necessary. The email should be formal and the student should provide specific information about themselves. The email should be worded similar to a professional cover letter and describe who the student is, when they plan to begin school, and some distinguishing information. This may include clubs, organizations, or past work experience. A resume should also be included. It is important that the email be tailored to each faculty member—this means that it should include specific information about why the student is interested in that particular faculty member's research and not be a generic letter that could be intended for anyone. Also, name dropping any past students or recommendations doesn't hurt.

What to Look for

Location

Many people choose a graduate school based on location. While this is a very important consideration for those who need to stay near work or family, it greatly limits the possibilities. Graduate students who have selected schools based on the location are usually limited to the research projects that are available in the department, and not necessarily what most interests them. If relocating is an option, it can broaden the opportunities that are available and better match the students' research interests.

Advisor

A compatible student–advisor relationship is essential to success in any graduate school. The advisor provides guidance throughout the degree and is a key player in facilitating graduation. An initial phone interview is a good way to get a feel for a faculty member's personality. This interview is not only for the faculty member to decide if they want to take on the student, but also for the student to decide if they want to work with that advisor.

One important consideration is the advisor's flexibility. This includes vacation time and working on weekends and holidays. At some schools, students on a research assistantship are not entitled to any vacation time—even during school breaks. However, some advisors are willing to negotiate vacation time and some are not. The amount of weekend work time varies from school to school and also from lab to lab. In some labs, equipment time is at a premium and in order to get experiments done, many students run them at night and on weekends. On the other hand, some labs are deserted on weekends. Again, this is a personal preference, but it is good to know a potential advisor's expectations for his or her students.

Another consideration is the advisor's involvement with his or her students. Some students have difficulty finding time to meet with their advisor, while other advisors specifically set aside time for these meetings. Additionally, some advisors prefer to be very involved in the student's work, while others choose to let the student work independently. It is a good idea to pinpoint how much help/supervision a student would like from an advisor and then choose the personality type with which they can work best.

Always ask about the advisor's research interests. This is beyond the research interests listed on the department's Web site. Even within a field like microbiology or food chemistry, there are many different areas in which to focus. Some may be interested in dairy, while others in meat. Perhaps the student is interested in statistics and would like an advisor who also finds that important. This is also a good opportunity to get a feel for their personality as they talk about their passion.

Ask about the advisor's other students. For example, what are they working on, what classes they are taking, how long it is taking them to graduate and finally, get contact information for one of those students. Contacting another student in the faculty member's lab is a great way to get a feel for how students interact with their advisor. It is also a good resource for information on campus life.

The Department

Food science departments vary from school to school; because of this, a careful consideration of a department's strengths and weaknesses should be a factor in choosing a school. Understanding the requirements and options for graduate students is key to picking the right department. These may include teaching responsibilities, required number of credits, specific courses, and a minor field of study.

At some schools, food science graduate students must have taken specific courses that provide the fundamentals of food science, and if they have not done this in their undergraduate degree, then they will need to take them as a graduate student. This is especially critical information for students who did not receive an undergraduate degree in food science—as they may need to complete all these courses during their graduate degree.

Different departments offer different graduate level courses; it is a good idea to check out what graduate courses are offered and when they are offered. Some courses may be offered every semester, every year, or even every other year. Knowing when these courses are offered and how often may help to narrow down schools based on the student's graduation timeline. It is also beneficial to look at the types of courses that are offered. For example, one school may offer a progression of food chemistry–based courses, while another may offer varied topics in food science. While one student may find the first option attractive, another may prefer the second. Finding out this information ahead of time may not only help in the decision making process, but also focus the degree curriculum. Additionally, some programs offer the option of obtaining a minor in another department. This may or may not be of interest, but is something to consider.

A quick way to get a feel for the department is to ask about their statistics. This includes information on their current students, alumni, and faculty. It may be useful to know how many students are enrolled, and how many graduate according to their expected timeline. Also find out the types of jobs alumni take after graduation. It may be useful to ask about the other faculty members: how many are there, what their specialties are, and how many teach students. Understanding the dynamics of the faculty is useful for graduate students as many need advice outside of their lab, or need to use other equipment or supplies during the course of research. Additionally, it is useful for a department to have a range of expertise—this can affect the types of graduate courses offered as well as the lab equipment/supplies available.

Assistantships

Many food science graduate students receive some sort of assistantship to fund their education. These assistantships are generally competitive, and it is best to apply early in order to obtain one. There are a number of ways faculty members can obtain money to fund a graduate student's education. Perhaps the most common is research grants—these are obtained after the faculty member or graduate student submits a grant proposal. Fellowships may also be available to students with strong academic achievement. Additionally, teaching assistantships are also offered.

When looking for financial assistance, the student should determine how funds will be made available—research, teaching, or academic achievement. If the money comes from a grant, it may also be important to know if the faculty member is applying for funds or if the student will be expected to apply for grants to get funding for a project. For teaching assistantships (TA), the student should determine how many courses they will teach and how much time per week they will be expected to devote to that course. It may also be important to know which courses they will teach and what the responsibilities are. For example, the TA may grade papers, lead a discussion or lab, or teach the entire course. Not all assistantships are alike; some assistantships pay a stipend, some pay tuition, and some pay both. Be sure to ask what costs are covered by the assistantship. This could be a deciding factor when considering a school.

If offered an assistantship, the student should understand what will be required of them. Some students work on one project throughout their degree, and that project is ultimately their thesis. However, other students will work on many projects, which may or may not contribute to their thesis. Determine if the faculty member knows what the project is ahead of time, or at least has an idea of the nature of the project. Knowing at least the basics may help in making the decision based on projects that are most interesting. For example, a vegetarian may not be interested in working on a project that will involve working on chicken. Also the nature of the research may impact the type of instrumentation that will be used in the project. The student should be comfortable with the equipment, or at least willing to learn to use it. Additionally, assistantships vary in the amount of time students are required to work and the amount of vacation they are allowed.

A final consideration is how long the assistantship will be offered. Some assistantships may only cover one year, while others can go on for many years. Asking in advance can reduce the shock factor of discovering that funding has ended or run out and it is now the student's responsibility for funding the remainder of the degree. Also ask what will happen to the assistantship if the graduation date exceeds the initial expected timeline. Some faculty are able to extend the assistantship if the student takes longer, while others stop paying immediately after the expected time frame ends.

Graduate School Visits

Visiting the graduate school is a great way to get a feel for the faculty member's personality, the department, the other students, and the campus as a whole. Because this takes up time and resources of both the student and the faculty member, this visit should only be to schools under serious consideration. The graduate school visit should answer any final questions the student may have.

On the visit, the student should talk to the potential advisor's other students—preferably away from the faculty member to get their honest and unbiased answers. The faculty member may arrange this for you, but be sure to request it if he/she has not. It is also beneficial to talk to the other students in the department about their experiences.

If it is possible, ask to meet with other faculty members. They will share their opinions of the department, the courses, and also talk about their research interests. They may also provide some interesting stories about the faculty member you are visiting. Visiting with another member of the faculty can also provide more insight into the strength of the faculty in that department.

Finally, take a campus tour. Graduate students don't generally get around the campus much, but it is nice to see what is out there. This can also provide insight into the types of students on campus and activities outside of class. Prospective students may be interested in the student unions, sporting teams, or recreational sports and this can offer some perspective of what is available. For students who are interested in campus life outside the food science department, it may also be useful to ask the other food science students how often they get around the campus and what types of activities they participate in.

MS or PhD

The choice between a Masters and Doctoral degree is both a personal and professional decision. It is based on career goals, interest in research, and desired time it takes to complete the degree. It may also be based on what research projects are available.

Careers in food science require varying levels of education. For some, a Master's degree is sufficient and a Doctoral degree may make a person overqualified for the position. However, others (such as a professor) necessitate a Doctoral degree. Examination of the degrees held by current employees in the desired department or area of work or the requirement section of a job description are good places to seek out degree requirements.

Personality also drives the decision between a Masters and PhD. Those who really enjoy doing research are well suited for a Doctoral degree. Additionally, some may want to spend years exploring the science and conducting research. However, others may prefer to work in an industry setting in contrast to academia and choose a Masters degree.

Finally, the research assistantship may determine the degree. The amount of work and research required for a project distinguishes it as either MS or PhD level. Typically, a MS project is well structured and can be completed in about 2 years. Doctoral projects typically require more initial research or experiments, are less structured and take longer. Some schools require that a MS degree be obtained before a PhD, but others allow students to go directly from an undergraduate degree to a PhD program.

Summary

Selecting the graduate program that best fits a student's interests and goals can be difficult. Graduate programs in food science are not a "one size fits all" degree and it is important to consider both personal and professional factors when making this decision. Prospective students should consider the faculty members, school, department, and campus as a whole to make the right choice.

Chapter 11
The Transition from Undergraduate to Graduate Student

Kelsey Ryan

Introduction

Graduate school is a new and different endeavor when compared to undergraduate studies. The differences primarily lie in the fact that the graduate student will have an independent research project that he or she is responsible for. Beginning graduate school can be a challenge since there is a lot to learn about research, there are courses to take, and the social aspect of graduate school can be very different. Two major things that tend to make graduate school difficult for some is the often unstructured environment and the absence of direction on how you should allocate your time. There is really no formula for graduate school success, and the advice given in this chapter is from my own and other current graduate students' experiences, so choosing what you think will work for you is most important.

By the time you have come to your department as a graduate student, you will have likely chosen a major professor to work with and hopefully this advisor will help you adjust to graduate school. Your major professor will likely be an integral member in the unofficial team you will build to help you through graduate school. One of the first things your major professor will probably help you with is determining the structure of the food science graduate program you are in. Most programs involve a combination of taking courses, doing research, and writing a thesis or dissertation on the research, and perhaps being a teaching assistant. The coursework required will vary between universities but will probably require competency in basic food science courses. In addition to doing research under your major professor, you will likely have a graduate committee of faculty members inside and outside your department to whom you will defend your thesis or dissertation.

K. Ryan
Purdue University, West Lafayette, IN, USA

R.W. Hartel, C.P. Klawitter (eds.), *Careers in Food Science*,
DOI: 10.1007/978-0-387-77391-9_11, © Springer Science+Business Media, LLC 2008

Getting started in graduate school can be difficult since it is hard to know what the expectations are, how to start a project, and how to deal with setbacks, for example. This book chapter serves to give some pointers from current food science graduate students who can help you out when transitioning from undergraduate studies to graduate studies in food science.

Expectations of a Graduate Student

There are many stated and unstated expectations of a graduate student, from setting goals to meeting deadlines. One important expectation of a graduate student is to take personal responsibility for the direction of your graduate study. It is also important for the student to establish and maintain a good relationship with the major professor and utilize other faculty and departmental resources. This section will focus mainly on two of these major expectations of graduate students—personal expectations and expectations of the student to have a good working relationship with his or her major professor.

First, when making the transition from undergraduate studies to graduate studies in food science, it is vital to remember that this is "your" journey. It is encouraged that you set goals for yourself and take charge of your needs. Goal setting should be a priority when you get to your department, if not before, and it is advised that you discuss your goals with your major professor so you are both on the same page. Some reasonable examples of the types of goals you may consider are when you aspire to complete your degree, what organizations you want to get involved with, and the research techniques in which you would like to be an expert. Although you may base some of your goals off of other current and previous graduate students, when considering the goals you want to achieve, it is important to avoid comparing yourself exactly to other graduate students. One current food science graduate student emphasizes that all graduate students' experiences are different due to the nature of research and customized course schedules, and as a result, progress is perceived differently. Everybody accomplishes their research and educational goals in their own way, so the key is to figure out what will work for you, set practical goals, and work diligently to achieve those goals. Also, share your goals with somebody else so you have a supporter to encourage you to complete your goals. In addition to simply setting goals, when you accomplish a goal, be certain to celebrate your achievements and be proud of what you have done, because it will motivate you to continue to reach your further goals.

Along with setting goals, another personal responsibility in graduate school is to schedule your time well. As an undergraduate student, most of your responsibilities likely revolved around courses. In addition to courses, as a graduate student, you will have research, writing, and perhaps an obligation as a teaching assistant, among other things inside and outside of school. Some graduate students recommend that you schedule time for these activities along with some time for personal obligations such as health and wellness, friends, and family since graduate school is not worth sacrificing your personal commitments and hobbies over. Alternatively, some students prefer a varied schedule that changes often and keeps them on their toes. How you go about scheduling your time is often up to you. Many major professors are open to letting the student choose his or her own schedule, but make sure to discuss working times with your major professor. Decide early in your program if you are willing to work nights and weekends, or work only during business hours, and plan your experiments around your personal preferences. Developing a consistent schedule may help keep you on task and help you avoid underworking or overworking, but a varied schedule works well for some students too.

No matter how you decide to schedule your time, during graduate school you will probably have more on your agenda than courses like when you were an undergraduate. The upside is that you will likely have a major influence in the decision of what courses you take. Many graduate programs actually have relatively few required courses, but recommend you take a certain number of credits related to your research. Being able to choose more courses as a graduate student again lets you organize your schedule the way you want. Also, influencing course decisions is one of the great perks of graduate school because you will, for the most part, be able to design your course load so that it relates to your research, and so that it is in line with your own interests too, and will help you in the lab and when writing. Additionally, there may be time to take courses less related to food science, like business or leadership courses, for example, if that is what you desire.

In addition to having a hand in selecting the courses you take in graduate school, you will also be able to select your graduate committee. It may be hard early in graduate school to know who you want on your committee, but make an effort to learn about the faculty inside your department and in other departments throughout your university so you can make good decisions about which professors to select for your committee. Talk to other graduate students and members of your lab for suggestions on who would be best suited for your committee as far as research interests, expertise, and personality goes.

Setting and accomplishing goals academically and nonacademically, doing research, and taking courses is very feasible in graduate school, but it is very important to stay motivated in graduate school to accomplish everything. There is no perfect way to find motivation that works for everyone, but if staying motivated means taking a break during the day, taking some time off, or changing your daily schedule, do what it takes, within reason, to get motivated. It is easy to get burned out if effort is not made to find what it is that invigorates you. If you are motivated you will be excited about your research and course material, and you will likely be more consistent and effi-cient at accomplishing goals. Not only can motivation come from yourself, but it can also come from your peers, your lab members, and your major professor. Motivation from your major professor may come in different ways, but ideally your major professor will inform you of what you are excelling at and perhaps what other routes you may be able to take to motivate you to achieve your research goals. However, your major professor may not always simply tell you these things and it may take asking for regular, informal reviews from him or her to find what it is that you can work on next. It is also important to review your professor too and let him or her know what you need, and ask for help if you do not know how to bring about changes. When asking for feedback from your major professor, take it seriously and use it as encouragement to accomplish more in your research. Also, asking for reviews may mean receiving criticism, but hopefully it is constructive criticism. Since criticism is inevitable in any field, one must learn quickly to cope with it in an appropriate manner, remember not to take it too hard, and try to use it as motivation to do better.

Next, graduate school in food science is focused a lot on research. It is an expectation of the graduate student that he or she learns methods in the lab, and is fluent enough to teach them to other students. Not only must you know how to perform your methods, you also need to know the methods in and out. When asked by interviewers, faculty, or students about the methods, you use you must be able to accurately describe them in detail. Graduate students should also be experienced in designing and managing experiments and meeting project deadlines. There are a plethora of deadlines that graduate students must meet at the lab, department, course, and university levels. It is the graduate student's, not the major professor's, responsibility to meet deadlines. It is also the graduate student's responsibility to abide by rules of the university, department, and lab.

In addition to taking responsibility for personal roles in graduate school, the graduate student must also maintain a good working relationship with his or her major professor. First, the role of the major professor or advisor is to guide and evaluate, monitor student progress in coursework and research, and discuss research goals and expectations regularly with the student. Ideally the major professor will be a mentor to the student. If this is not the case, it is a great idea to seek out somebody who can be a mentor during graduate school. This mentor should be willing to give you feedback and answer questions you have. The major professor should also make sure the student shows understanding of the research being performed and relevant material. Additionally, a good professor will help you organize your schedule by providing rough guidelines on how to balance coursework, research, writing, and perhaps teaching, and also understand the need for freedom and a personal life while in graduate school. If your major professor does not provide some guidelines, ask for assistance from him or her, or again, seek out another person who can lend you a hand.

Also related to the relationship between the student and major professor, it is the student's responsibility to stay in contact with the major professor and preferably have regular meetings scheduled in order to discuss research progress and goals. Several graduate students suggest that it is important to figure out how to best communicate with your major professor and maintain frequent communication, even if it is not accomplished in face-to-face meetings. All major professors are different—some are in the lab often, some are available to talk to students any time, and some professors are very hands off and do not enter the lab often. Therefore, not only is it good for the student to determine how to communicate with his or her major professor, it is also a good idea to determine the frequency of meetings or correspondence. Some students desire more guidance and direction than others, so letting your major professor know this right away is smart. Guidance and direction do not necessarily mean, however, that the major professor gives you all of the answers and tells you what to do. Finding a good medium between excessive attention and negligence will be beneficial. The goal again is for you to learn the necessary information and techniques to be successful in your career, so a major professor who gives all of the answers is not really helping after all. Determining communication styles between you and your major professor will help to ensure an interactive relationship that will ultimately benefit you.

Finally, university and departmental resources are available to help graduate students. Get to know other faculty, staff, undergraduate students, and graduate students in your department. Having a good departmental network will be beneficial when you need resources such as equipment and advice.

Project Development and Project Management

Project development and project management are tools that graduate students will be very familiar with before the completion of graduate school. These are undoubtedly key skills important in any job that a graduate student will master. Different people will have different advice about project development and management, so you must choose what advice to take and what advice to pass up, tailoring your project development and management to your own needs. This section centers primarily on organization and time management, the major professor's role in research projects, and research preparation.

As mentioned earlier, keeping an organized schedule can be very beneficial in graduate school and several current food science graduate students made note of this when surveyed. When it comes to project development and project management, organization and time management become even more important. Many graduate students recommend recording deadlines in a calendar so you have a timeline to follow throughout your graduate studies and so you can plan ahead for important events. Also, making a calendar of goals may help you stay on track for completing them. Again, do not compare your personal deadlines and goals to other graduate students because every student will go about accomplishing things differently, so do what you need to do to meet your goals and your personal graduation goal. One graduate student highlighted that at any point in graduate school there will be things you can be doing that will contribute toward your goals and your personal graduation deadline. These activities may not correspond exactly with your calendar, but always try to keep yourself busy at work so you keep yourself on track.

In addition to making and working toward deadlines in graduate school, there are efforts that can be made to help your research go smoothly. Just as it is important to be organized as a graduate student in general, it is also important to be organized in the lab. Keeping a lab notebook, and updating it regularly, will help you stay on course in the lab. Also, careful planning of experiments will benefit you since you will be less likely to have looked over important information or have to repeat experiments, as one graduate student brought to my attention. Students who plan well will likely meet their graduation goals.

Developing and planning experiments will become easier as graduate school progresses. To begin, it is important to read and comprehend the literature related to your research early so you know what you are doing when working in the lab, so you understand what your research will contribute to the literature, and so you will know which methods to use. Additionally, some current graduate students recommend keeping notes on the literature you have read as this may help you understand it better and will likely help you in the future when writing. Keeping notes will also help you determine what literature you have already read so time is not wasted in re-reading literature that is not relevant. Keeping notes on your completed research is important, but it is also important to keep track of ideas for future research. Having these ideas and conclusions written down will help you when writing manuscripts or your thesis since you may not always remember the exact conclusions, or all of the conclusions, you thought of earlier. Developing and managing a research project may sound daunting, but jumping right in and trying things out in the lab is the best way to learn. Everybody will make mistakes doing research, so it is best to get started right away so you can learn from your mistakes early on.

Before you actually start performing your research experiments, reading and understanding the literature related to your research, as mentioned several times earlier, will help you throughout your research project. Start reading early, the day you start graduate school, and spend a lot of the first months reading new material. Also discuss the literature with others in your lab to enhance comprehension. If you do not understand the literature it is really important to ask for help in grasping the concepts. It is important to read the literature, but it is also important to be selective. If you are unsure of exactly what to be reading at first, ask your major professor or a lab member what material will be most relevant to you. Several graduate students also recommend writing the literature review for your thesis early on. This works for many, but some research project paths change throughout graduate school, so the literature review will likewise change. This decision of whether to start the literature review early must be made by you and your major professor and both approaches can work well. Also, many graduate students suggest that it is easiest to write up your materials and methods as you are doing your research since it is hard to remember exactly what you did later when writing, especially if you do not keep up on your lab notebook.

When managing a research project it is good for the student to stay in contact regularly with his or her major professor. If the student or major professor deems it necessary, the student can run research plans by the major professor before executing them. Working closely with the major professor will likely lower the incidence of developmental and methodological mistakes. Also, the student should consult his or her graduate committee for guidance on next steps in research and methods. These faculty members will probably have expertise in different areas than your major professor, so their input may be very helpful in planning research. Also, these are probably the same people that will decide if you graduate, so it is wise to get their opinions on the direction in which the research should go early in the project.

Critical to managing any work project is remembering to stay balanced inside of work and with life in general. Even though a lot has been said in this chapter about setting goals and achieving your graduation target, it is still important to be steady and take it easy sometimes. Setting a daily schedule that is realistic and filled with activities outside of the lab or school can help a student stay on task and stay motivated. Scheduling activities outside of work stimulates some graduate students to be more productive at work and encourages students to spend their time at work wisely. Some research projects can be incredibly overwhelming for graduate students and in these situations it may help to divide the project up into parts and finish each part individually before moving on to the next. This way the project will seem less intimidating and it will be easier to get started. Overall it is important to have research scheduled and organized so you can complete your degree in a reasonable amount of time, but it is more important that you do not overwork and become stressed out.

Part of your schedule will be filled with courses, and these will often dictate when you can do your research. Many times, courses will not be offered at the best times of the day for your research, so they are one more event to consider when organizing your schedule and managing your research. Many students and faculty recommend taking most courses early in the timeline of graduate school versus late. Although it is good for students to jump into research right away, many students do not have research projects or are not ready to start research for a few months. This allows these students to take several courses early since it is hard to focus on coursework when busy with lab work. Taking courses early on allows the student to focus more on research toward the end of the graduate degree and not be interrupted by courses that are not offered at ideal times.

How to Deal with Research Setbacks

Setbacks in research are truly unavoidable. Sometimes no matter how great you plan, your research will surprise you. It is hard for anyone to take setbacks easily, but there are some important things to consider when your research does not go as expected. It is important not to get discouraged; have a positive attitude, and look at setbacks as beneficial.

First, do not take research setbacks personally. No matter how organized you are and how much you plan, research is hardly ever successful the first time and it will take a lot of hard work and practice to get methods working, to set parameters, and to get valid results. Things also often take longer than expected, so be patient. It is important not to get discouraged and to make sure to seek help and feedback when things do not go as planned. When research does not work out at first, slow down and think about what went wrong and what your next steps can be. This is a good opportunity for you to discuss research with your major professor. Your major professor will likely have different and innovative ideas for what steps to take next since they have been doing research longer. Your major professor should be able to help you think of alternate solutions to research problems, not tell you what to do next. Other good places to look for help with research are other members of your lab or other graduate students. You never know who has had experience in your area of research or with your methods, so it is wise to ask around for guidance. Also, as mentioned before, the graduate committee is a source of advice and feedback. Perhaps a committee member is experienced in the area of research at hand and can steer you in a better direction. Also, look for feedback outside of your university. Other researchers may have experienced your problems before and may be able to help you. But be sure to run your plans by your major professor first before consulting another person for research advice.

When something does not go right in the lab, or when it does, make sure to document exactly what happened in your lab notebook or elsewhere. In the future, this information could be very important for helping you or others working on the same research. Also write down ideas in your lab notebook since these may help you when you repeat your experiments later or, as mentioned earlier, when it is time to write. The most important thing to do when you encounter research setbacks is not to become discouraged. Eventually you will begin to feel like an expert in your area of research. It takes time and ambition to complete research, so be perseverant.

In fact, instead of getting frustrated over research setbacks, try to do the opposite and stay positive. Having a good attitude will help you take setbacks easier and help you be more motivated to continue researching and finding the answers. Since research setbacks will likely happen to every researcher, it may be a good idea to plan for setbacks and always have an alternate plan in mind. It is a good idea to talk about alternate research projects other than the primary research projects when you and your major professor discuss your research. If you have other projects in mind when the primary research project does not go as planned or cannot be implemented, it will be a lot less stressful to decide on the next steps.

When things do not go right in the lab, it may also help to take a short hiatus from your experiment. Working on courses for a day or two, or working on other projects, may help take your mind off of the problem and may help you be more efficient the next time you work on the experiment or when it is time to decide what to do next. Also, taking a break may allow for time to think more deeply about the research. In this time, you may realize that your research is not really going down the right course. In this regard, research setbacks may be beneficial because they may help you realize that you need to take a new direction with your research. Again, discussing the research with your major professor is probably the best first step to take. Then you and your major professor can decide if the current track is where you want your research to really go, what to do next, and if it is best to start on something new. Encountering research setbacks can also be good because it will force you to reflect on your research and learn more about what you are doing since you may need to revisit the literature and look to the basics to find some answers. In the end, research setbacks can provide the opportunity for you to think critically and come up with new, perhaps significantly different, ideas for the research project that may take it in a new and better direction.

Unfortunately, the research goals for the major professor and student do not always coincide. If the research is not progressing in the desired way for the student, or if there is in general major disagreement between the graduate student and major professor, the student can usually change projects or major professors. Obviously this is not the most desirable outcome for a graduate student, but it is better to get out of a research project or research partnership with the major professor early versus late before more problems arise and before time is wasted.

Other Skills Needed for Succeeding in Grad School

When you think of graduate school, developing a social network may or may not come to mind. Yes, the purpose of graduate school in food science is to attain greater knowledge in the field by taking courses and doing a significant amount of research, but it is also important to meet new people and build relationships. Graduate school should be a complete experience and being social contributes to the experience. This means that a graduate student should not spend every night and weekend in the lab, but should take time for old friends and for meeting new friends. When activities, related or not related to school, occur inside or outside the department or university, graduate students should try to participate. Many current graduate students believe that graduate school is lot more enjoyable if you have friends to share your experiences with and activities to participate in.

Also essential in graduate school is working on professional development and on building a network. Professional development reaches beyond typical food science classroom learning and encompasses skills such as development of clear and consistent technical and nontechnical writing skills, effective organization and time management, and learning about topics other than your area of research. A good way to continue to work on professional development, and networking as well, by continuing your education and by meeting new people, is to attend professional meetings and conferences and become involved in professional associations.

Next, networking receives a lot of attention in all fields, and it is really important to start building your network early. A network requires effort and work to build and maintain, but the benefits are too many to be measured. Many students and professionals find their jobs by using their network for assistance and networking can start right in your own department. Make an effort to seek out faculty, staff, and other students in the department. An easy place to start is with your own lab members. You may never know the great things about the people you work with everyday unless you ask them about themselves. Learn about what the new people you meet are doing and learn about their past experiences in food science. You will learn something new about a person and make a friend, and perhaps in the future you will be able to utilize their expertise for help on a project or when searching for a job. If you are not comfortable approaching new people on your own, ask a lab member or your major professor to help you get started in meeting new people within and outside of the department. It is important not to be shy though—you want to meet as many people as is comfortable and get your name out there. Also be sure to keep a good rapport with everybody you encounter because, as many people say, the food industry is small, and you never know who you may be working with or for someday.

Not only should you make an effort to network with people at your level and above your level, it is also good to get to know less experienced graduate students and undergraduates in your department. Great friendships can ensue and you may have the opportunity to mentor an undergraduate by sharing knowledge and answering questions, just as you will likely have a mentor as a graduate student. Befriending an undergraduate or new graduate student is beneficial to both the undergraduate and the more senior graduate student. The less experienced student will become part of your network and you will be able to help him or her out by talking about the undergraduate program, graduate school, or finding a job for instance.

A great way to meet new friends, develop professionally, and build your network is to become involved in a club or an organization. Many food science students choose to participate in departmental food science clubs and the Institute of Food Technologists Student Association (IFTSA), but being involved in any organization of your interest will benefit you now and in the future. Now you will have the opportunity to participate in meetings, competitions, and events, and in the future you have a network to rely upon. Participating in educational events outside of class and research will help you learn even more and give you the opportunity to apply skills. Belonging to an organization or association will allow you to focus on something outside of school, perhaps motivating you to be more effective when working. Being involved in organizations also gives you the opportunity to develop leadership skills. Running for an officer position in any club will give you leadership experience and help develop your professional skills. Participating in extracurricular activities will appeal to future employers as well because it shows you are willing to do many different types of things.

Despite the appeal of involvement in associations like IFTSA and others for professional development, sometimes graduate students spend too much time focusing on activities related to these organizations and associations. Faculty members have observed graduate students spending so much time on extracurricular, professional development activities that they cannot finish their work. This is a situation where balancing a schedule and obligations and having exceptional time management skills are essential. Participation in outside activities can be very enriching, but making sure time is managed so graduate work can get done is paramount.

Summary

In summary, this chapter has focused on how to be successful in transitioning from undergraduate studies to graduate studies in food science. First, there are expectations of graduate students within and outside of school, and many revolve around taking personal responsibility for your studies and communicating with others. Second, it is important to be organized in your research, take time to read the literature, plan experiments in advance, and in general, try to stay organized and focused while in the lab, but also have fun outside of work. Next, dealing with research setbacks can be difficult, but it is important to avoid frustration and to take time to reflect on problems with research. Finally, in graduate school, networking and getting involved with organizations or associations can help advance your career and make graduate school in food science enjoyable while meeting new people and learning new skills.

Chapter 12
Faculty Expectations of Graduate Students

Richard W. Hartel

When looking for a new student a few years ago, I considered an international student who wasn't available for me to interview personally—something I've come to require before I accept a student into my research group. After some preliminary discussion, I asked her my "behavioral" questions by email to give her an opportunity to provide me with some insight into her qualifications and character. I asked her to describe experiences where she had to resolve a conflict with someone else, where she had faced and overcome a hurdle, and to describe her motivation for graduate school. In her response, which started by noting a particular interaction she had had with her father, she presented me with a well-written documentary of her skills, into which her responses to my three questions were woven. Being the sort of person myself who would have bullet-pointed a response and detailed specific activities to document those skills, I was greatly impressed with her ability to think more broadly than my specific request, yet get at the heart of my questions in a creative approach. I accepted her as a student immediately because those are the attributes in a graduate student I value most highly.

Her response convinced me that she could handle the rigors of an independent graduate research program. That is, she would be able to deal with the usual setbacks in research in a creative way and, most importantly, that she was highly motivated to succeed on her own while still being willing to learn from me. And, she did this with a personal touch that struck a chord with me. However, other professors may not have taken her response the same way I did—we are all different and look for different things in our students.

R.W. Hartel
University of Wisconsin, Madison, WI, USA

R.W. Hartel, C.P. Klawitter (eds.), *Careers in Food Science*,
DOI: 10.1007/978-0-387-77391-9_12, © Springer Science+Business Media, LLC 2008

All Faculty are Different

Each faculty member has his or her own expectations of graduate students. Mine, as noted above, relate to motivation, creativity, independence, and the ability to resolve problems and conflicts. I look for someone who will work within my system of mentoring because I've learned over the years that if I accept students who do not fit within my system, we both have huge problems. Although I've only had to let one student go from my program (out of 46), there have been several students who have required what I considered to be too much of my time and energy, and given back less than I expected in terms of research productivity. To be fair to those students, they didn't have a picnic with it either since I had trouble giving them what they needed. The importance of both professor and student in understanding the needs and approaches of the other person cannot be stressed enough. Without a good student–professor relationship, graduate school can be a horrible experience, for both people.

And each faculty member is different, so it's difficult to generalize things completely. At one extreme are professors who give daily guidance to each graduate student, essentially telling the student what to do. Other professors either do not have the time for daily interaction or expect that students will be able to work by themselves on a daily basis. It is important that the needs of the graduate student match those of the professor, otherwise there will be continual battles—battles that students generally cannot win.

I think differences among faculty arise, at least in part, from their outlook on what a graduate program should entail. At one end of the spectrum are faculty members who consider graduate students to be the vehicle of their own research success. Since high-quality research publications mark faculty success, professors are interested in generating substantial amounts of high-quality data for such publications. These professors tend to guide every detailed step of what the student does to ensure it meets their high standards. Unfortunately, students in such a system are generally not given the opportunity to expand and think for themselves, since they are carrying out someone else's detailed guidelines. Some students thrive in such an environment, where they are given detailed guidance at every step of the way. Although perhaps this is an easier route for students (being told what to do), this approach typically does not foster independent research skills in the student, something that is a highly desirable outcome of graduate school.

On the other end of the spectrum are faculty members who never seem to have time for individual graduate students and essentially leave them unguided for long periods of time. This lack of direct contact may be forced due to other commitments (i.e., department Chair/Head responsibilities, excessive teaching or advising load, substantial extension activities, etc.) or may simply be due to a philosophy that students develop best when forced to manage on their own.

In my own personal experience, I had a PhD advisor who quit midway through my research (he found selling real estate to be a more lucrative occupation), leaving three graduate students in the lab to fend for ourselves. Our department found a replacement advisor to oversee our project, but the three of us had to train this new faculty member in our research topic while still trying to get useful guidance from him about completing our research. When I was hired after graduation, the company noted that being able to complete my PhD under those circumstances showed that I was clearly capable of independent work. So in the long run, it was a good experience for me, although in the middle of it there were many frustrating days and much grumbling about how "life is not fair" (the comment made by our advisor as he left).

Ideally, a faculty member would recognize the needs of each individual student and provide exactly what they need when they need it. Each student is different and needs a different amount and type of guidance and their needs may change over the course of a program. A faculty member who can provide the support needed, at the appropriate time, and then back away to let the students perform (and even struggle) on their own is doing the best to help students develop their individual research skills. But, this is a skill that professors do not learn in school. Professors develop their own style over the years by trial and error, often starting with the model experienced in their own graduate programs. Like new parents learning how to be a parent (or new teachers learning how to teach), advising graduate students is something that professors must sort out for themselves.

Finding the Right Professor

Most graduate students have some amount of control over their selection of faculty advisor, and it is highly preferable that students select their advisor based on that person's reputation in dealing with graduate students. Having a good working relationship with your advisor is just as important as working on a topic of interest to you or getting a degree from a certain school. Incoming graduate students should honestly assess their needs in an advisor (e.g., do you need more guidance or less), and then find a person who will help them make the most of their skills.

Some questions you might consider include:

- How much freedom will he/she give you in your research project? Will they require that you do as you are told or encourage you to make the project your own?
- What expectations does the professor have for his/her students? For example, is there a policy on vacation time or does he/she have specific requirements for time spent in the lab?
- Are there any unwritten performance expectations?
- How much time will the professor spend with you and what will be the nature of that interaction?
- Are there regular group meetings, and if so, what happens at these meetings?
- How does the professor express a personal interest in your success? All professors will say they are concerned with student development, but each professor expresses that interest in different ways.
- Does the professor respect his or her students, and how is that respect demonstrated?

As a prospective graduate student looking at different programs and faculty, interview each faculty member to make sure that his or her approach works for you. Also interview the students in the lab, since they are the ones who really know what that faculty member is like. And then carefully consider the responses. If a student says the professor is good to work with, find out exactly why, so you can tell if their reasoning matches your personal needs. If a student says a professor is not good to work with, again find out exactly why, because that students reasoning might not be the same as yours (their mentoring needs might be different than yours).

Typical Faculty Expectations

Once a student is accepted into a professor's research lab, there are certain expectations of that student. Again, there are variations among faculty members, but for the most part, all faculty want students who are hard workers, can think independently, and who generate a lot of high-quality data. The following points are my personal perspective on what faculty look for in students, mixed with some advice about how to make the most of your graduate experience.

Work ethic. I've heard of professors who expect graduate students to be in the lab every day before they arrive and leave after they leave at the end of the day. Since most faculty work 8–10 hours a day, graduate students are expected to put in long hours. Other professors are not so concerned about time spent in the lab, especially now when Internet opportunities mean productive work can be done from almost anywhere. However, all faculty members expect to see high levels of productivity, regardless of where that work gets done. A student who doesn't produce results is in danger of being put on probation and eventually being let go. All universities have policies regarding "satisfactory progress" and students can be let go if they do not meet these expectations. Fortunately for students, the process of letting a student go for lack of progress is formalized so that they get a fair hearing.

However, a hard worker is more than someone who puts in a lot of hours. What about those hours between classes or experiments, when there is nothing urgent to accomplish? How do you spend that time? I've seen students surf the Internet, "chat" with friends online, do the crossword puzzle, and even play computer games because there was "nothing to do." Don't think your professor doesn't know when you're not focused on research (or classes)—he/she might not say anything at the time (although some professors have been known to rant and rave about how students spend their time), but you can be certain this will be reflected in how he/she supports your search for a job when you graduate. Remember, your major professor will be the primary person you turn to when the time for letters of recommendation comes.

A hard worker fills time in productive ways. Got some extra time? How about studying the literature, perhaps even getting a start on the Literature Review chapter? Good researchers scour the literature for anything and everything that's been done before that is related to their research area (and explore very widely at the start of your project). Before you start any project, you should do a complete literature search and carefully read anything and everything related to your project. Don't just take your professor's word about why your project is a good research topic—learn it for yourself. Make it your own personal project by adding your own insight based on a literature survey and add your own opinions about what needs to be done. Furthermore, consider submitting a literature review for publication, something that benefits both student and professor.

How about exploring other fields to see how other scientists approach problems similar to the one you're working on? You could even explore what others in your lab or other labs in your department and beyond, are doing. Can you come up with a valuable side project to work on during the "lulls" in your main project? If you run out of options, feel free to ask your professor for a side project to keep you busy. He/she will be thrilled with your work ethic and will undoubtedly find side projects that enhance your skills and background. These activities require internal drive and motivation, but are highly valuable in many ways (learning, experience, good recommendation letters, etc.).

Internal drive. Graduate school is so much different than undergraduate school. As an undergraduate, courses are generally structured to provide enough details to complete assignments and tasks. Research doesn't work that way. Research requires an intrinsic motivation and innate curiosity to get up each morning and get into the lab to run experiments (or to the computer if you're modeling). It's easy as a graduate student to say "I'll do that tomorrow" because deadlines are less structured than in undergraduate class obligations. The drive to get to the office or lab and collect some data has to come from within the student. And again, if there are "extra" hours in the day, find some useful/beneficial way to spend that time.

Motivation. Many years ago, I accepted a student into my lab because he seemed like a good guy, had a good background (decent GPA and excellent interns) in a particular area of my research, and came from a good undergraduate school. For some reason, however, I missed the point that he wanted to work with me because his girlfriend had taken a job in the area and he wanted to be close to her. In hindsight, after a year or so with him in my program, it became evident that his main motivation to be in my program was to be near his girlfriend (now wife) and not to work on a graduate degree. At the start, he did not have the drive and passion for research that I expect in a student. Another time, I took an international student into a PhD program without a personal interview because he challenged my questioning him about his motivation. He bristled at the thought that a graduate student at this level would not be motivated. I took him instantly, and he finished a PhD in 2 years, whereas the student who came to graduate school to be near his girlfriend took 3 years to finish a MS degree. Although the MS student ultimately did some very good work, the process to get to that point was often challenging for both of us. Motivation to do graduate work is now one of my highest priorities when evaluating which students to take into my lab.

Natural curiosity. Research is about exploring new ideas and principles. By nature, a scientist should have an exceptional level of curiosity to learn new things, explore new concepts, and develop new knowledge. How do you advance your level of curiosity? How can you become more curious about things when that's not your nature? Here are several tips (www.lifestyl.com/index.php/education/how-to-study/how-to-develop-curiosity/).

- Keep an open mind
- Don't take things for granted
- Ask questions relentlessly
- Don't label something as boring
- See learning as something fun
- Read diverse kinds of reading.

Typically, creativity in research (or science in general) means reading everything and anything, both within the field and in other fields, which might relate to the research topic. Note that this approach also helps with the next important graduate student trait, knowing how to get past research roadblocks. And can help fill in "lulls" in your research in highly productive ways.

Creative solutions to research roadblocks. Perhaps the main attribute that separates good researchers from others is their ability to deal with experimental setbacks. There will always be times when methods don't work right, equipment breaks down, or something happens to stall a research project. How you approach things when this happens is an important characteristic. Perhaps your first response is to run into your professor's office to explain the problem and ask what to do. Fight that urge! I've kicked students out of my office for asking me what to do, with the advice that they come back after they can pose five different possibilities for us to discuss. My aim is to get the student to think through the problem him- or herself and come up with multiple creative solutions. At that point, I'm happy to discuss the problem with the student and undoubtedly we'll come up with even more solutions (or maybe just approaches to get to solutions).

Where do you come up with ideas and potential solutions to get around research roadblocks? That's where the creative part comes into play. Reading widely in the literature, both within the field and outside, is one way to find ways that others have found solutions to certain types of problems. Expanding your thinking and knowledge, beyond foods, can often lead to creative ways to resolve problems in your own research (sometimes creativity is simply taking a tool or concept from another field and applying it into your own field). Another approach is to learn every detail of how equipment works, especially in cases when equipment malfunctions provide roadblocks. However, each situation requires its own unique creative approaches.

You might consider developing your creativity through reading books, attending creativity workshops, or taking creativity courses (either online or at your university). Although many of these books, courses, and workshops are very general, some of the approaches they use to developing creativity may work for you.

Attention to detail and being thorough. Good, reproducible research is done by people who pay close attention to every little detail of what they're doing. People who haphazardly approach their work are much more likely to get variable results that are difficult to interpret or don't prove/disprove their hypotheses. While developing a method or technique for your research, you should think about every detailed step to consider if a variation in that step will influence the outcome. How does the time taken for each step affect the results? Does the temperature in the lab affect your methods, and thereby affect your results? What other influencing factors must be controlled to minimize variability?

For many years, I had two experienced researchers in my lab who had developed superb lab skills. They knew almost intuitively which parameters were going to have a significant impact on the results. In one study, they figured out that our results were being affected whenever a truck drove past the building on the road outside our lab. The variability caused by the vibrations from trucks passing outside was sufficient to make the results meaningless, and it wasn't until we installed a vibration table (for damping motion) that we could make sense of our results. Developing a careful approach to your studies will go a long way in optimizing your research results, or at least getting the most out of your methods.

Being a thorough researcher means making sure you "dot all the i's" and "cross all the t's" at every step of the way. From doing a complete and extensive literature search to checking out each and every assumption, pay attention to details at every step of your project. A thorough researcher stands out from the crowd.

Know how to get what you need from busy people. It's an important skill to know how best to approach your very busy research advisor to get the input and feedback you need to do your job and make satisfactory research progress. This is a skill that will translate very well into a future career, especially in industry where you will need to interact regularly with your busy supervisor. Even though it's your professor's responsibility to give you feedback and guidance, the daily pressure of deadlines (class preparations, proposal due dates, meeting preparations, etc.) often means there isn't enough time in a day to meet with every graduate student. Scheduling weekly meetings may be a good approach, although flexibility will still be needed since travel pressures mean he/she might not always be there for your regularly scheduled meetings. But suppose you're really stuck or, on the other hand, have a great inspiration that you need to discuss first with your advisor—how do you get into their schedule to get their time?

The skill of knowing how and when to knock on your supervisor's door is an important one to master. One of my least favorite phrases when a student appears at my door to get some of my time is "are you busy?" Of course I'm busy. I work 10–12 hours per day because that's what it takes to get all the work done that I need to get done (and even then I'm usually behind). From writing papers and proposals, to preparing for classes or advising students, there are more things to do than there are hours in the day—that's true for every faculty member. However, unless your professor is up against a critical deadline (like getting ready for class in the next 15 min.), he/she will usually be willing to spend some time talking, as long as you approach them in a way that allows them to put down what they're doing and attend to your needs. Develop a pleasant tone of voice and politely ask for some of their time—that will go a long way to getting the time and input you need from busy people.

Also, recognize that at the moment of your interruption, he/she is focused over a task that probably doesn't involve your research question. Help him/her switch gears to now think about your situation by clearly yet concisely setting the stage for your question or inspiration. A few minutes of setting the stage properly (emphasis on brief and concise) will go a long way in getting your needs met.

Satisfactory Progress

Each institution has regulations for what constitutes "satisfactory progress" of a graduate student. Often, it simply comes down to the assessment of the major professor to decide whether a student is making good progress or not, although in some cases, a faculty committee may be involved in directing a student's program and evaluating whether satisfactory progress toward the degree is being made. Either way, each student is expected to make satisfactory progress in both coursework and research. Satisfactory progress in coursework is relatively easy to assess. The student must have a coursework plan in place that will allow them to complete all course requirements in a reasonable time and attain a grade of B or better in all courses. Satisfactory progress in research, however, is not always so easy to define.

From a faculty perspective, satisfactory progress in research typically means the student is spending adequate (and productive) time on their research project and generating enough meaningful data to fulfill the needs of the funding agency. Meeting the expectations of the funding agency with significant, usually publishable, results is certainly one critical outcome of satisfactory progress. However, some projects are more difficult than others, particularly if they require developing new techniques, and a student may spend months futilely trying to develop methods. In this case, it's sometimes a judgment call as to whether the student is making satisfactory progress since there is no hard data as the output. The "output" may simply be that a certain method or idea doesn't work. But, at what point is that decision made, and who makes that decision? The student might say the challenges are unreasonable, whereas the faculty might say the student is not capable of completing the task. There is no simple answer here, which is why the student–professor relationship is critical.

Another aspect of satisfactory progress is the professional development of the student, especially as an independent researcher. All faculty members expect that students will improve their research skills: developing hypotheses, performing literature searches, statistically designing experiments, performing experiments with due care to reduce variability among replicates, analyzing results using appropriate statistical methods, drawing conclusions, and proposing future work. Typically students enter a graduate program with only rudimentary skills in these areas and are expected to become self-sufficient, independent researchers when they leave. Yet, there is a balance between student development and research progress. Funding agencies typically don't care about student development—results are all they care about. Thus, faculty members often are faced with the dilemma of supporting student development even when research results are not forthcoming. Again, a good student–professor relationship is critical to make the best of an undesirable situation.

Summary

In my opinion, the student–professor relationship is probably the single most important factor that will determine the success of a student's graduate experience (and beyond). For a faculty member, taking a new graduate student is a risky thing. There is a commitment made for 2–5 years (MS or PhD) that is not easy to get out of. If a student doesn't match a professor's mentoring style or for some reason is not performing at a satisfactory level, there are options for releasing the student from the lab. However, the process of firing a graduate student is not easy (and it shouldn't be, to be fair to the student) and the loss of research productivity when a wrong choice is made can be devastating. For the student, it is critical that the advisor provide satisfactory guidance. A lack of adequate directions and guidance from the professor, particularly at the start of a MS program, can lead to a very disappointing graduate student experience.

It's important to recognize that much of a student's success, both during the graduate program and when seeking a job, is dictated to a large extent by the major professor. It is important that the professor provide the necessary guidance during the graduate program; however, it is the student's responsibility to meet the needs of the professor since a strong letter of recommendation from the major professor is an essential part of getting a good job upon graduation. As a student, think carefully about how you meet the faculty expectations discussed earlier since these are likely to be the primary basis for that letter of recommendation.

Careful consideration on both sides is needed to ensure that the graduate experience is beneficial for both student and professor.

Part IV
A Successful Industry Career

Chapter 13
A Successful Industry Career

Moira McGrath

You've completed your BS, MS, and/or PhD in Food Science. Congratulations! Now, how do you get that dream job in the city and state in which you want to live? This chapter will cover how to do just that with tips on how to write a good resume and cover letter, develop excellent interviewing skills, learn how to look for that perfect job, and negotiate a fair package for yourself.

Resumes

Your resume is your foot in the door. It is the only way to get an interview, so it must be informative and well written. If your resume is poor, it is a poor reflection on you. You must "sell" the reader as to why they should call you for the interview vs. another student with the same degree. There are two basic types of resumes, one for industry, one for academia or government. For industry-based resumes, use this basic outline.

Name, address, phone number, and email address. Don't confuse the reader with too much information. Make it simple.

Maryann Student
1181 Park Avenue
Elizabeth, NJ 07208
908 555-1212 (cell)
maryannstudent@uccl.com

Please note, as a student, you might consider including both your college address and your home address (parents) so that if someone is trying

M. McGrath
OPUS International, Inc., Deerfield Beach, FL, USA

R.W. Hartel, C.P. Klawitter (eds.), *Careers in Food Science*,
DOI: 10.1007/978-0-387-77391-9_13, © Springer Science+Business Media, LLC 2008

to get in touch with you after you have graduated, they will know where to find you. If you are using a cell phone number as your contact number, and plan on continuing to use that cell phone after graduation, this might not be necessary.

Objective. The objective tells the reader what you want to do in your career. For example, *Seeking a position that enables me to utilize my MS degree in food science in the field of new product development.* This is especially important when you are a student with little to no history of employment. If there is no *Objective*, the reader has no idea what field you want to pursue. For example, let's assume that the job opening for which you are applying is to work in *new product development* for a cultured dairy products company (yogurt, cottage cheese, etc.). The job description mentions that there will be a lot of microbiology work. If you have an interest in microbiology, the *objective* is a perfect place on the resume for you to mention it. Note the difference between "Seeking a technical position in the food industry" vs. "Seeking a position in the field of new product development that will allow me to utilize my degree in food science and my interest in microbiology." The second objective grabs the reader's attention.

Education: List your most recent degree first, and then list the rest in chronological order. Summer courses or short courses should be listed after all of the college degrees.

PhD Food Science, University of Minnesota, 2008
MS Food Science, South Dakota State 2006
BS Chemical Engineering, Cornell University 2003
"Sanitation Management" 6 week course Summer, 2001

You can add information about your thesis topic, but be brief. You don't need more than a title for your research. Your GPA can be listed if it is above a 3.0. Don't include high school; it is no longer of interest, even if you attended a fine preparatory school. If you attended courses that might be of interest for the company, include that information here. For example, if you attended a summer program learning culinary skills, that should be included under *Education*. However, don't include education that does not apply to the position you are seeking. A summer class in rock climbing does not belong on your resume.

Work Experience: This section describes, with dates, what work you have done in your career.

Summer, 2007 Kraft Foods, Inc. Chicago, IL
Intern

- *Conducted work on bottled tea beverages, focusing on new flavors and improving antioxidant levels.*
- *Brewed innovative tea and protein beverages for various tests.*

"Bullet-point" your experience under each company. Begin each point with an action verb—past tense when appropriate. Imagine that the reader has no idea what food science is. Explain clearly what your responsibilities were, and what you have accomplished.

Include brand names of products whenever possible. There's nothing like name recognition to pique another person's interest.

Awards and Achievements: If you have won awards, scholarships, competitions—list them here. For example:

IFT Product Development Competition, Captain, University of Nowhere team

Won First Place for "Low Moisture Hot Dogs" 2006

Personal Information and Activities: If you have been involved in sports, clubs, teams, other academic departments, competitions—list them here. Whether it is in college or outside of school, (local fun runs, garden club), include it. Age, salaries, religion, and marital status should not be on a res-ume. Remember, companies hire *people*, not resumes. Make yourself human.

Captain, University of Nowhere Swim Team, 2005–2007
Ran New York City Marathon, 2006

Limit your resume to two pages. If you have fewer than 4 or 5 years' job experience, a one-page document will be sufficient. Lists of publications, patents, and presentations are not part of the resume. These are separate documents. Names of references should not be included on your resume, but have a list prepared and provide them to your recruiter or human resources contact when requested.

Overall, ensure that your resume is attractive and easy to read. Use an interesting, modern typeface, and leave adequate margins. Make sure there are no spelling errors or grammatical errors. If English is not your first language, have a friend or colleague whose first language is English read and correct it.

When emailing your resume, make sure your document is transmitted in a form that can be immediately opened and printed by the recipient. A MS Word Doc or PDF files are best.

Sample Resume

Maryann Student
1181 Park Avenue
Elizabeth, NJ 07208
908 555-1212 (cell)
maryannstudent@uccl.edu

OBJECTIVE

To use my food science experience in a leadership role in the development of new, innovative food products for consumers to enjoy.

EDUCATION

University of Nowhere 2005–Present
Masters of Science
Major: Food Science Specialty: Product Development & Research
Graduation Date: August 2007 with GPA 3.7

LeftOut State University
Bachelors of Science 2005
Major: Food Science Minor: Packaging Science GPA 3.0

WORK EXPERIENCE

ABC Ice Cream Company **2005–Present**
ABC is a student-run food production facility.
Manager

Responsibilities included:

- Currently manage 11 undergraduate students.
- Plan, manage, and execute projects ($500– $25,000+)
- Hire new employees and manage their progress
- Negotiate directly with suppliers for ingredient needs and work with a national food service merchandiser as a main customer through direct interaction and local community retailer (contractual value $1,500– $25,000)
- Implemented new labels for pints and ½ gallon containers.
- Organize ice cream socials and other functions for organizations on/off campus

XYZ Creamery **2002–2004**
Technician
Responsibilities included:

- Production of ice cream and delivery of product to retail locations around campus
- Complete sanitation cleaning and mechanical assembly of processing equipment
- Prepare media and conduct microbiological testing for quality assurance
- Inventory management through record keeping of weekly production and sales

M&M Dressings, Inc. **Summer 2004**
M&M products are all natural and preservative-free salad dressings, sauces, and dips
Research and Development Intern
Responsibilities included:

- Responsible for coordinating R&D projects for new products (salad dressings and sauces), product improvements and product matching
- Collaborated with others through product ideation and problem sol-ving
- Conducted several product audits, along with supplier audits and plant tour
- Evaluated products through quality assessment; moisture, ph, visco-sity, acid and salt concentrations.

HONORS & AWARDS

- Graduate Student Teaching Assistant Merit Award recipient – May, 2007
- National Collegiate Dairy Products Judging Contest – Second place (Oct. 2005)
- Regional Collegiate Dairy Products Judging Contest – Fourth place (Oct. 2004)
- Air Force Academy and West Point Academy Nominee – 2001

INTERESTS AND ACTIVITIES

- Actively involved in the Culinary Initiative team at ABC, where we collaborate with other companies to develop innovated products.
- Judged the 2006 National Ice Cream Retailers Association Ice Cream Contest for various flavors.

- Assistant Coach of the Dairy Product Evaluation Team—Food Science Department
- Involved with the first annual "Culinology" collaboration with IFT and RCA at the 2006 IFT Food Expo.
- Team captain on the Food Product Development team
- Participated in collegiate athletic programs: swimming and tennis
- Currently an active member in the Institute of Food Technologists and Institute of Packaging Professionals

Cover Letters

Do I need a cover letter and what should it say?

A well-written résumé is, in just about every case, the most important document in a candidate's professional portfolio. Cover letters, specifically tailored to each query, are the next most-effective tools.

You must include a cover letter every time you mail, fax, or email your resume to a potential employer. The purpose of the cover letter is to make the reader so interested in you that after reading your resume, he or she will call you immediately to learn even more. Your cover letter is the appropriate place to let your personal self shine through. You may have talents, abilities, or interests that would be inappropriate to mention in the resume proper, but which a company deciding whom to interview would consider major assets.

For example, if your preferred geographic location is Boston, be sure to note that in letters to Boston-based employers. If you want to work in an area outside your realm of direct experience, the cover letter is the place to make a compelling case for your specific knowledge and inherent abilities that will allow you to succeed.

Directed enthusiasm might also be effective in your cover letter. Have you been inspired to create a new use for, or application of, the employer's product? Let the hiring authority know!

Temper that enthusiasm with wisdom and brevity, however.

Appearances Count

Limit your letter to, at most, two or three carefully constructed paragraphs. As in your resume, be definitely, absolutely sure there are no spelling or grammar errors. Proofread, proofread, proofread, then have another person do the same. Trust your instincts. If you're not sure about the spelling of a word or construction of a phrase, it's probably wrong. Don't gamble that it won't be noticed!

The physical appearance of your cover letter is almost as important as its content. Of course, if you're mailing it to the employer, it should be on the same stationery as your resume. Be sure to leave attractive margins and to type your name below your signature. When faxing, don't use textured or toned stationery for your source document.

Whom to Address and How

If you're answering an ad that calls for you to respond to a company's human resources department, call the company and get the human resource manager's name. Address your letter to that person, being sure to include his or her title.

If you know a contact's name but not the specific title, ask. That small effort just might set you apart from the rest of the crowd. It certainly will give you a better idea of how to gear your letter and maybe, even, your resume.

Yet another reason to know your audience is that, armed with the name, you will be able to avoid the Mr./Ms. dilemma. Be sure to use Dr. wherever appropriate; avoid Mrs. That designation is reserved for social and certain other occasions when a woman is being addressed in her capacity as a wife or widow: Mrs. John Doe.

Your cover letter is your marketing tool, your door opener, your spokes piece. It's worth time and energy to make sure it represents you in the best possible light.

Remember that you have but "one shot" at impressing a hiring authority, who may spend as little as nine seconds scanning all the paperwork you have worked so hard to create. It's essential that you get every possible bit of mileage from each document you present.

What to Say and What to Avoid

Mention specifically the title of the position for which you are applying, and note the source of your information. If a mutual colleague recommended that you apply, be sure to mention that colleague's name.

Don't waste time and paper describing yourself in trite generalities, no matter how glowing.

Shortlist three or four examples of the training and/or experience that make you a perfect fit for the position at hand.

Interviewing Skills

What are They Looking For?

You have been called by a major food company and been invited in for an interview. But as a recent graduate, what do you have to offer? You haven't really *done* anything yet! But you offer more than you think. When you interview for your first job, remember that those who are interviewing you KNOW that you have limited to no industry experience. So, relax, don't try to compete with those who have more experience than you. The interviewers at your entry-level job interview are not looking for technical skills; they are looking for "soft skills."

Interviewing Necessities

You have heard the expression "first impressions count." This could not be truer than in a job interview. If you look sloppy, have bad manners, or come in to the interview unprepared, even if you are the "best qualified" candidate, kiss that job goodbye, because you are not going to get it. Absolute MUSTS include:

Professionalism

Always dress in a professional manner (suit and tie for men, business suit for women)—even if your interview is on a Saturday morning in Florida.

Turn off your cell phone before the interview (If by some horrible accident it rings during the interview, don't answer it!)

Advise the interviewer/hiring authority promptly if your flight is delayed or you're stuck in traffic.

Greet everyone you meet at an employing company with a firm handshake. Look the interviewer in the eye. Be gracious to everyone you meet, including security guards and maintenance personnel.

Refrain from calling an interviewer by his or her first name unless invited to do so. Address PhDs as "doctor" unless told otherwise.

Refrain from any use of profanity, even if the "corporate culture" seems to allow it.

Answer all questions—including salary questions—directly, even if you've been asked exactly the same question by three previous interviewers.

Treat the human resources associate with the same respect you've shown the hiring manager.

Provide requested names and phone numbers of references in a timely manner.

Return messages promptly, whether from your recruiter or possible new employer.

If you are unable to talk when you receive a phone call, inform the caller about when you will be available.

When you are leaving a message for a hiring authority (or anyone else, for that matter), always say your name clearly and your phone number slowly. (It's helpful to say "My phone number is . . ." so the recipient is prepared to write it down.)

Make sure your cell or home phone message is clear and businesslike. Your phone message starting with "Yo!" is not good business practice. Check for phone messages regularly in case someone is trying to call you for an interview.

Instruct roommates about how to respond to business calls to your home.

Finding a new position is a job in itself, but a poised, professional candidate will have an immediate advantage over those who are less prepared. Don't sabotage your search—good manners make good business.

Personality

While grades are important, they are not nearly as important as well roundedness. Students who have been on teams and competitions (i.e., Institute of Food Technologists (IFT) Student Product Development Competition, College Bowl, etc.) are perceived as having a sense of team spirit, fairness, an ability to communicate, and a competitive nature; all traits that work very well in industry. Since industry is very team driven, an individual who "plays well with others" is considered a top candidate. A good attitude and an outgoing personality go a long way.

Communication skills are critical. Making technical presentations is only one type of communication skill. Others include knowing how to get your point across in a small meeting, explaining a technical problem to a nontechnical person, getting things done with and for people who aren't at the same level as you are. Communication skills also means speaking English properly and this applies to both, English as the first language student and the international student. If English is your second language, you *must* master the language. Practice, practice, practice. Take courses in eliminating your accent. Have your friends correct you. If others are unable to understand you, they won't hire you.

If English is your first language, you must know how to express yourself properly as well. Language peppered with "like" or "um," or "ya know" is an absolute no–no. Giggling throughout an interview is not appropriate.

All students should take a public speaking course, either in school, or courses taught by companies such as "Toastmaster." Taking these courses can really help you speak clearly and effectively, abilities you will use for the rest of your life.

Good Work Ethic is rarely discussed, but is an underlying and extremely important trait. You were probably raised being told about doing the "right thing." "Finish the job," no matter how small or trivial. "Do the job right," not sloppily or half-way. "Treat others as you would have them treat you." Those rules still apply, particularly in the work place. The candidate who has mediocre grades, but a good strong work ethic will be the candidate who gets the job over the candidate without the work ethic.

Internships The most important thing you can do while in school is to work in an internship or two. Students who have worked in industry during the summer or a semester definitely have an added advantage. Why? If an interviewing company sees a resume of a student who worked two summers, it tells them that, first, you have an interest in working in industry (vs. academia). Second, you have some practical experience, which brings you to the top of the heap of resumes sitting on the interviewer's desk. The benefit to the candidate is that now he or she has one or two different companies to approach after graduation, which already know their work and work ethic. It is common for a company to make a full-time employment offer to those who have worked summer internships with them. The company has had an opportunity to learn about you as well as you have had an opportunity to learn about them. It will also give you practical experience so that you can understand what jobs might be available to you when you graduate. You may find out that, as an example, you have a passion for food safety, something you may not have known had you not worked in the field for the summer.

Research the company Learn about the company, either on the Internet, or by speaking to other scientists who work there, (or both). You will impress your interviewer by having knowledge about the company and its direction. Are they introducing a new line of cheeses that are expected to blow the competition away? You need to know this before your interview. Be informed.

Leadership skills are considered a "plus," so if you are captain of your team, make sure to include this on your resume, and talk about your leadership abilities and experience in your interview.

What to Expect

When you are invited to a company for an onsite interview, you'll be greeted at the door, brought in to sit down in someone's office, and will meet with several people from a hiring team. You may be there for a few hours, or all day. You will be asked a series of questions, sometimes in "group" interviews, or one-on-one. The purpose of this exercise is twofold; first is to confirm that you have the right technical qualifications for the job.

Second, and most important, is to see if you are the right "fit" for the company. All companies have a personality, or "culture." The interview team looks at each candidate to make sure their personality matches the company's culture. Cultures can vary from being aggressive, marketing driven, and fast paced, to research driven with a slow but steady approach to growth. Many times the candidate who does not have the "perfect fit" in technical skills is the candidate who gets the job because they fit the culture. This obviously works in your favor as well, as you want to work in a culture where you feel comfortable. Imagine that you are a very detail-oriented type individual and you accept a position in a company where you are surrounded by peers who only see the big picture and are not interested in the details. This would not be a good long-term relationship. You would drive each other crazy!

Where Do I Find a Job in Food Science?

So you've done the best you can do in school, your resume and cover letter are both prepared, and you have practiced your new interviewing skills. Now it's time to look for a job. Where do you start? To whom do you send your resume? You have a few options.

Companies Interviewing on Campus

There are many companies who send representatives to college campuses to interview graduating students. These companies generally look for the "cream of the crop" candidates. One Fortune 50 company sends representatives to the chemical engineering departments of only the top schools and offers positions to *juniors* to lock them in for a start date immediately after graduation. But companies sending representatives to interview food science students is happening less and less, probably due to the small sizes of classes, and the expense of doing so without any guarantees of success. This is a good option for students if it is available at your school, but it is not always available.

Internet

There are a few ways of searching for jobs online. You can send your resume to a particular company's Web site, as many companies list their job openings. Go to the Web site of the company you like, and look at their job listings. If there is a posting that you feel fits your background, send your resume. You may not get an acknowledgement that the resume was received, but it probably was. You can also send your resume to a blind ad, (no company name listed) or post your resume on job boards like Career Builder or Monster.com. These are all passive ways of finding a job, and there are a few problems with these strategies. As a student with little to no experience, your resume is now in a huge pool of resumes of other students who are looking for a job just like you. Also, with blind ad and job boards, you really don't know where your resume went. Who were those "blind" companies? Who has a copy of your resume? You may have sent your resume to companies who were "fishing" for resumes. These can include both reputable and disreputable companies that are looking to see "who's out there" without having a job opening—companies who are "fishing." When you are looking for position, whether you are employed or unemployed, be discreet, and be careful. It is not recommended to send resumes without knowing to whom you are sending it. If you want to use the Internet in your job search, make sure that you have a person's name and company name to which to send your resume. Know what the job requirements are, and make sure that you meet those qualifications. Refer to the job title in your cover letter, and explain to the reader in your cover letter why you qualify for the job. But searching for a job on the Internet is probably the least effective way to job search.

Recruiters who specialize in placing food scientists are another option. Sometimes hiring companies will ask an executive search firm to screen the candidates if they have an entry level position open. But this does not happen often, as the recruiter charges the company a fee for the service. The candidate has to be much better qualified than all the other candidates, as hiring this candidate will be more expensive. Again, this is an option, but certainly not the best.

Networking

Networking is by far the best approach to a job search, and this will be true throughout your career. Eighty percent of all jobs filled in the United States, in food science as well as all other industries, are filled as a result of networking with others. Use every opportunity to network at food science clubs, events, and social gatherings. Join the IFT Student Association and get involved with their programs so that you can network with industry leaders. Go to the local IFT events and talk to people who work for companies where you might have an interest in working. Go to the IFT Annual Meeting, and become a volunteer so that you can meet others in industry. Companies hire people whom they know and like. Get out there and get to know people who can help you in your job search.

Thank You Notes

You just had a terrific interview with a company you would love to work for. What next?

Once upon a time (this is a true story), a job candidate felt that a personal interview had gone exceptionally well. He expected an employment offer immediately. After 2 weeks, however, no offer was forthcoming, and the candidate's enthusiasm turned to dismay. Only then, when he was mentally reviewing (for the hundredth time) everything that had been said—and done—before, during, and after the interview, did he remember that he hadn't sent a thank you note.

Hastening to correct the error, he express mailed a letter to the employer who, it happened, had never before received such a communication. The candidate's thoughtfulness so impressed the employer that she phoned him and made the offer on the spot.

It's essential to send a thank you letter, whether or not you are interested in a given position. A letter or, at least, a note keeps the door open for future contact. It is a polite way to let the company know where you stand just as quickly as you would like to know their thoughts. It sets you apart from less conscientious candidates and proves you are caring as well as courteous.

It is the perfect venue to reinforce your enthusiasm for the job and the company, and the logical place to remind the employer of your qualifications and to summarize the strengths you can bring to the position.

A thank you note also reminds a hiring manager that he or she needs to respond to you.

Determining where to send your letter shouldn't be difficult. But is one letter enough? The hiring manager will, of course, be the principal recipient. But don't forget the human resources staff who set up the appointment and made travel arrangements or a particular secretary or receptionist who contributed to your well-being during the interview process. Let them know you appreciated their help.

A thank you letter may be handwritten on quality stationery and snail or express mailed. Avoid email unless the employer has given you his or her personal email address *and* has indicated that messages are checked frequently.

Deciding what to say isn't as difficult as saying it succinctly—and sincerely. Thank the employer for the interview, express your interest in the position, and clarify points that might have gotten short shrift during your meeting. Don't neglect to summarize why you are right for the job, personally (e.g., you want to live in that part of the country) as well as professionally.

While fairly simple and fun to write, thank you letters are essential to any job search. The right letter could clinch the offer.

Negotiating a Fair Package

Let's assume that you have been made an offer or two. How do you know which job is the best one for you? What leverage do you have to negotiate a better package? How will you know if there is any room for negotiation?

First, ask yourself: "Is this the job I want, in the city I want? Does it use my best skills as well as give me an opportunity to grow in my career?" If the answer is "no," tell the company right away that you are not interested, and thank them for their time. Be gracious, as the food science industry is a small world, and you don't want to shut any doors for future consideration. If the answer is yes, then let's evaluate the offer fairly. An offer includes:

Salary

Companies generally have a salary range that they offer students based on the students' level of education and industry experience. (Another reason to do those internships!) Unfortunately, as a student, you don't have much leverage to negotiate. You should research what the company has offered other students who have a similar background to yours, and make sure that it matches the offer you received.

Benefits (health, dental, retirement)

Good questions to ask would be "When do the benefits start?" (the first day of work or in 6 months?) and "How much, if any, do I have to pay for the benefits?"

Bonus

Is there one, and on what is it based? Personal performance? Company performance? When is it paid? Quarterly, annually? Bonuses can add from several hundred to several thousand dollars to your annual income.

Authorization to Work in the US

This is an issue that has been more and more difficult to overcome. Are you authorized to work in the United States on a full-time basis, and on what basis are you authorized to work? All companies realize that our undergraduate and graduate programs in the United States are filled with excellent international students. But not all companies have the financial capability to sponsor. It is an expensive process, and, other than the very large food companies, many cannot justify this expense. Also, the time involved in sponsoring someone is enormous. The company has an obligation by the US Government to post the position for a certain amount of time before they can open the opportunity to candidates from outside the United States. This is, again, why internships are so important, as a company may be willing to "fight" for a candidate they already know and in whom they have confidence. Companies that are international in scope will be more open minded to sponsorship than a US-based company whose sales are predominantly US driven. If you are not authorized to work in the United States, and a company is willing to hire and sponsor you, you must be clear on what the terms and conditions are of your sponsorship before you accept the offer.

Relocation

Does your offer include moving your household goods and/or your car? Are they reimbursing your expenses, or is it a lump sum? Any of these options are good; it depends on your personal situation. If you are a married student with children, you might need a little more help than a student who just has a suitcase and a car to move.

Start Date

Are you ready to go to work right after graduation, or were you expecting to take the summer off? You need to be flexible. . .don't expect a company to wait more than a few weeks for you once you have graduated.

Asking for More

You were offered your dream job, but there are a few things more that you need. How and when do you ask? The time to negotiate a better package is *before* you accept the job. Don't wait until after you have the offer letter and have verbally accepted before you "remember" one more thing you need. However, if you negotiate a better package, and the company gives you what you asked for, it is not appropriate to turn the offer down. If you are negotiating, it is assumed that in good faith you will accept the job if you are given the extra dollars/benefits, etc. that you have requested. Let's use the example that you have been offered a position where the salary and bonus is fair, but the company offered you only $1000 to relocate yourself to the new city, and you have a boat that needs to be moved. You have researched the price of the boat move, and found that it will cost $1500. How do you ask for more without appearing greedy? You must first do your homework and find out exactly how much your move is going to cost. Get written estimates that you can provide to the new company. Tell the company how interested you are in this position, and their company, but ask if they will increase the relocation dollars to cover the written estimated expenses that include the boat. Remember that they may say "no", and you will then have a second decision to make. Do you want the job despite the fact that they did not cover your additional $1500 expense? There is no shame in taking less than you asked for. You are making a long-term decision, not a $1500 one. But the time to ask for more is *upfront*, when you are in your best bargaining position. Once you are an employee, it is too late.

Accepting the Offer

When you have been made an offer, you need to give your answer within a few days. It is not fair to any company to make them wait for your decision. Waiting for your next interview is not an excuse for stalling your answer for a few weeks. If you have done your homework, you will probably know *before* you have your interview whether or not you want to work for that company. If you like the people, the offer, and the company, don't wait. Accept the offer, and cancel the other interviews. If you are not sure, perhaps it isn't the right fit for you. Listen to your "gut." Do what your heart tells you is the right thing.

Summary

Taking the right steps while in college can lead to having a successful industry career in the field of food science. Obviously good grades count, but working in the food industry while in school (summer internships) is the best way to guarantee that you will be at the top of the list of candidates for the top jobs. Professionalism, good communication skills, and the right attitude during the interview process are equally important. Everything you do or don't do during the interview process counts. There are no short cuts to finding the right job for you. It takes work, just like anything else you really want in life. Some say looking for a job is a full-time job. That statement is not far from the truth. Writing a good resume and cover letter, and doing your homework before your interviews is a lot of work. But it can be very rewarding when you find the right company and job. Remember, it's all up to you.

Chapter 14
Employer Expectations/Managing Corporate Life

Dennis Lonergan

Have you ever wondered if you will end up working for a large corporation or an entrepreneurial business? Some people seem to thrive in one environment, but falter in another. Or, they just seem to enjoy one more than the other, while being able to function equally well in either. This chapter will not provide you a crystal ball to look into and predict your future career path. However, hopefully, it will provide you with information on life in the corporate world that you will find useful at some point in your career, be it in a large corporation or a small entrepreneurial venture, or both.

What You Bring to the Company

The short answer to this topic is your *knowledge*, *enthusiasm,* and *innovative thinking*. Or put in other words, "yourself." Rather than being a flippant answer to this topic, it really is something for you to think about in depth. By "yourself" I mean to suggest that you have had different knowledge and experiences than anyone else in the company. Hopefully, this will give you the ability to look at problems and challenges just a little differently than will anyone else in the company. You need to bring this unique set of skills and perspectives with you to work everyday, and most importantly, use them! Don't fall into "group think!"

There is a hypothesis, and a historical perspective, that big inventions often come from someone who is not just new to the specific problem which is being tackled, but also new to the area in general. This is described in *The Art of Scientific Investigation*, by William Beveridge, a book that I recommend to all scientists, regardless of where they are in their careers.

D. Lonergan
The Sholl Group II, Inc., Eden Prairie, MN, USA

R.W. Hartel, C.P. Klawitter (eds.), *Careers in Food Science*,
DOI: 10.1007/978-0-387-77391-9_14, © Springer Science+Business Media, LLC 2008

Surprisingly, and probably counterintuitively, being new to the company actually gives you some advantages over your more experienced coworkers in this regard.

In addition to benefiting the company, there is also one big benefit to you. If you practice bringing all of yourself to work each day, there is a reasonable chance that you will actually continue to enjoy coming to work! This is especially important after the elation of the first big pay check wears off, which usually happens after the equally big mortgage, furniture, or car payment bill arrives in the mail.

There is of course a potential downside to this approach. What I have just suggested contrasts with the approach of figuring out how "the company" typically goes about getting things done and then fitting in with that culture. Figuring out the norms and adapting to them will certainly earn you praise as being a "quick learner," being described as someone who "comes up to speed quickly" and is a "real team player." There is nothing inherently wrong with being any of these, and of course, they have their obvious benefits, such as helping you keep your job and even getting a promotion some day.

I think it is possible to have it both ways. You can still bring your whole self to work and not be described as a "square peg in a round hole" or worse yet, "disruptive."

As in most things in life, balance is required. Perhaps it is described in the concept of emotional intelligence. The concept is discussed in the book *Emotional Intelligence: Why It Can Matter More Than IQ,* by Daniel Goleman. It is well worth reading.

Preparing for Your First Day of Work

Since this is your first day of work, you obviously have passed the interview process. That means that you have done your homework and know about the company, their brands, mission statement, and possibly even the names of your supervisor's children (if you are really good with Internet searches). All of this is good, with the possible exception of the last item. So what do you still need to do to get ready for the first day at work?

The answer may lie in what you need to leave behind. My experience with most people who enter the workforce after just completing their MS or PhD thesis is that they feel they clearly were hired because of that work. This is certainly understandable. It is the research that you poured your heart and soul into for the last several years. It was summarized in what will probably prove to be the largest written work of your life. And if that wasn't enough, you had to endure several hours of professors making certain that although they were "granting" you this degree, you realize that they are still much smarter than you.

Forget it! Your thesis showed that you can do research and that you know how to approach and solve technical problems. That's it. The chances that the specific content of the work had anything to do with your getting hired, or with what you will be doing in your new job, are very small. But don't worry; with time, and perhaps a few sessions with a psychotherapist, you'll get over it.

You may also need to leave behind the comfort of being "top dog." Often, the position of "senior" graduate student in a lab is real, powerful, and in the end, useful for the smooth functioning of a college research lab populated with students, be they undergraduates or graduate students. In your new job, you definitely will be the new kid on the street, so don't expect folks to follow your dictums with heads bowed.

So is there anything that you need to do to prepare for work? Here are a few suggestions:

Find out the dress code. No sense feeling even more self-conscious by wearing a business suit when your coworkers are in jeans. Conversely, you probably don't want to show up in a Hawaiian shirt or a tank top and shorts. I've never seen a major food company where that fits the definition of business casual. (If you find one, let me know.) It may actually help to speak to someone at the company. When I asked a relatively new employee what they wished someone had told them before she started, her response was how cold the air conditioning was set in the building. She wished she had brought along a sweater to her first day of work, even though it was summer. This is the type of information that can only come from speaking to a future coworker at the company.

Read the packet of benefits information. Expect to spend time with the human resources department. You will be filling out W2 forms, making decisions on various insurance coverage's, 401 K deductions, etc. It would be helpful if you read the material that the company has provided you ahead of time, as it will make this process go more quickly. Also, bring along your Social Security card and a photo ID.

Bring an open mind. You are bound to learn a lot in your new job. It most likely will involve products and processes that at best you have only read about before. So be ready to be a sponge and soak up this new knowledge.

Relax. Before you know it, it will be your second day, then your second week, and then your second year.

Entering the Corporate World: The Details

There are several things you will need to adjust to, which you didn't need to be concerned with before entering the corporate world. This includes attending more meetings on more topics than you thought could possibly happen. However, probably the most important detail is something called "corporate culture."

A simple definition of corporate culture is that it describes "the way things get done" in a corporation (and it varies from corporation to corporation). A more cynical definition is that it is internal politics or unwritten rules of a corporation.

If you are starting to work for a large corporation and you ask the human resources department about corporate culture, they will probably hand you a glossy multicolor handout on their corporate culture or values. It most likely will include affirmation of their respect for ethnic and religious diversity, that it is the people in the company that make it great, and the role that the company plays in the larger community. All excellent, but that is not the part of "corporate culture" that speaks to how things get done in the corporation, and which is important for your success, let alone survival, in the corporation.

It may be useful to give some examples. You will also find some useful general information on corporate culture by doing a quick Internet search on the topic.

Does Everyone Really Own the Final Objective of a Successful Project, or Does Work Get Done in Silos?

For example, you are working on a new product concept, and you rough out a quick financial analysis of the approximate delivered margin. Will this be greeted as a sign that you are proactive and take initiative (as well as indicating that you were awake during your finance class, even though it was at 7:45 a.m.), or will the finance group get in touch with your boss to make certain you get the message that in this company, financial analysis is done by the finance group and not research and dvelopment(R&D).

Communication Style

Is a Power Point presentation (with animation) expected for any "important" meeting or is the culture one of black and white handouts, even if the meeting is with the CEO? It is best to check before hand. Doing it the "wrong" way will be taken as an indication that you haven't "gotten up to speed" yet in your new job. Mess up twice and you will probably be headed for a 3-day seminar on "effective communication."

Understanding the accepted norms for communication up the chain of command is another important aspect of corporate culture. For example, is it OK to discuss information with your boss' boss before you have shared it with your boss? In many cultures, this is not acceptable, even though all parties will claim to have an "open door" policy and to believe in the value of clear, timely, candid, and honest communication.

Getting Started on Projects

Read! This is the best advice that I can give to anyone starting on a project. What to read falls into several categories. These include:

1. Intracompany reports related to your project or similar efforts in the past. These can be technical, marketing, consumer insights, or finance reports. This can eliminate time wasted in redoing work that has already been done. Unfortunately, this is not always practiced. I actually sat through a presentation by a research team that enthusiastically described a technology they had developed and were suggesting that the company protect it with a patent application. Great idea, except for the fact that another team in the company had come up with the same technology and had filed a patent application on it 18 months earlier. All of this was in the company's electronic notebook system and was available to anyone in R&D who could read.
2. *Scientific literature.* Although your team is probably certain that it is the first to be exploring this idea, the reality is that others have probably tried something similar before. It won't be described the same way, but the underlying technical challenges have probably been investigated before by someone, and the results published in the literature. I think you will find that Google Scholar and Entrez PubMed are excellent search engines.
3. General interest writings on innovation and new products. Examples include:

   The Tipping Point: How little things can make a big difference by Malcolm Gladwill 1999: Little Brown and Company
   The Innovator's Dilemma, by Clayton Christensen 1997: Harvard Business Press
   Blink; The power of thinking without thinking by Malcolm Gladwill 2005: Little Brown and Company

In addition to reading, there is one large difference between how you got things done on your thesis research project and how you will get those same things done in the corporate world. *The change is that you don't have to do everything yourself.* In fact, you had better not. Again, best to illustrate with examples.

1. Chemical analysis. You will be expected to submit samples to your company's analytical lab. There may be some very specific or simple tests,

such as the specific volume of a baked item or pH, respectively, that you will do in your own lab. But on the whole, analysis will be done by another group. Learning to trust results generated by someone you have never met may be a challenge at first.
2. Sensory analysis. Same as above.
3. Literature searches. Same as above.

This freedom from doing routine analysis and other tasks will free up a lot of time. Some of that time will be taken up by the aforementioned meetings. Hopefully, there will be some time left for you to think about how you will approach the challenges of your new project. This is addressed in the following section.

Project Management

Many corporations use a standardized process for project management. This often consists of phases or "gates" that a project passes through on its way from an idea to finally launching the product. With this gate system there may be formal gate meetings, where "key stakeholders" (a.k.a., department heads, also known as "the suits" before business casual became the norm), decide if the project moves forward, recycles to gather more data, or is killed.

Killing a project is something that I should say a few more words about before moving on. First, it is not necessarily a bad thing. Most ideas don't proceed all the way to a new product launch with a multimillion dollar launch budget. "Fail often and fail early" is actually a wise model for product development. The objective is to minimize the time and money that is spent on a project before it is killed. Second, don't take it as a sign that your career has been fatally tarnished if your project is killed. It has not, and life will go on, both yours and the corporation's.

In place of a project gate system, a less formal process of key milestones that need to be passed may be used. These processes are all helpful in making certain that important questions are addressed, and addressed in some semblance of a logical order.

The order in which one answers questions is very important, but cannot be totally addressed by a project management system. One of my mentors put it this way: "*do the last experiment first.*" What this saying suggests is that you should think about what is the most challenging item on the list of things that must be accomplished for the project to be successful. Identify that item, or items, and work on them first. Since you are the technical expert on the team, this means that it will fall to you to identify the technical hurdles, and identify which one will be the hardest to overcome.

This is not a trivial task. My experience is that many people have a difficult time with it, especially since there are many items that must occur for a project to be successful. That is, any one of them not being achieved will result in the project failing. However, on this list of things that must occur for the project to be successful, there are one or two items that have the highest odds of tripping up the project. The key is to identify these and *work on them first*!

This may sound obvious, but human nature can tend to push us to work on those problems that we can knock off easily. The benefit is that this will allow us to show progress on the project, which is seen as a sign that the project is on track, the team is working well together, and that work should continue. However, it can also serve to increase the amount of time and money that is spent on a project which is killed because that most difficult technical hurdle could not be surmounted.

So now you have identified the key hurdle that separates success from failure on the project. The fun work, which your education and scientific training has prepared you to tackle, now lies before you. This work may require that you find a new innovative solution to the problem posed in the "last experiment first" exercise.

I suggest reading an article by Paul D Trokhan on innovation. There are hundreds of books written on the subject of innovation, but few of the authors have been named as an inventor on over 200 US patents. The article is entitled "An Inventor's Personal Principles of Innovation" and can be found on this link: http://www.allbusiness.com/technology/4508086-1.html.

In this article, the author outlines the fundamental principles that guided his research career, which covered over 30 years at Proctor & Gamble. Read them! He goes on to suggest that each of us develop our own list of principles. Here are mine. I hope you find one or two that are helpful in your work too.

Lonergan's Principles for Scientific Innovation

- A problem well defined in fundamental terms is 80% solved.
- The best person for the job is the one who knows what to ignore.
- What a fool sees and believes, a wise man reasons away.
- Don't read all the literature in the area before attempting to solve a problem. It will result in looking at the problem the same way as everyone else has in the past.
- When trying to stop an unwanted event from occurring, think about what you'd do if you were trying to make it happen.

- The least useful seven words in research are "we tried that before, it won't work."
- Have passion and set audacious goals. Then see who wants to follow you. They are the ones you want as colleagues.
- Be careful of falling in love with your own theories. Even Nobel Prize winner Niels Bohr said that his theory on the structure of the atom, although useful, was probably incorrect (and it was).
- If you don't look forward to coming in to work in the morning, figure out why and do something about it. Great inventions never came from someone who didn't like what they were doing.

Working in Teams

This seems to be an area that is currently being addressed in a student's academic career much better than it was a generation ago. Working on a project as a member of a team is now fairly common in many college classes. That experience is certainly helpful as you start your corporate career.

However, the part that will differ in corporate life is that the team is now much more diverse than was the team, for example, in your food engineering class. In that class, all of your team members were obviously enrolled in the same course, had taken similar prerequisite courses, and were at a similar point in their academic career. The team in the corporate world will probably be a cross-functional team. This means that there will be people from finance, sales, and marketing on the team, and they took very different courses during college that you did. The team will also be much more diverse in terms of demographics. There will be people like your self, just starting their career, and there may be people with 30 years of experience on the team, and people any place in between.

This diversity will lead to a much more robust answer to challenges the team faces, but it can also lead to misunderstandings and even distrust among team members. This will test your skills at building a strong team. There are many good books on team dynamics and how to build a strong team. One that I have found to be particularly helpful is *Managing for Excellence: The Guide to Developing High Performance in Contemporary Organizations* by Bradford and Cohen (Willey Management Classic).

Moving up the Corporate Ladder

Probably the first question that you will face in this area is "which ladder." In most large food companies there will two career ladders, one "technical" and the other "managerial." There of course will be company literature that describes the job expectations of each career path. In general, the technical path focuses on developing technical expertise in one or several areas, being an individual contributor as well as a cross-functional team player, and providing technical leadership. The managerial path will entail accountability for achieving business objectives, developing strategies to meet business objectives, and probably most importantly, skills to manage people and help with their career development.

However, I think there is one question that you can ask of yourself that will help with this decision. That question is: "will I be happiest being the person actually coming up with the innovative solutions, or will I be happiest being the leader of a group, and watching others in that group come up with the exciting inventions?" A reflective and honest answer to this question will help to point you in the right direction.

So, once you decide which ladder you want to ascend, how do you go about getting promoted (and promoted, and promoted, etc.). My suggestion is that the way you move up the corporate ladder is by *not* focusing on it when you are new in your job. Doing excellent work and accomplishing more than is expected of you is the foundation for moving up the corporate ladder.

One path to doing more than is expected is captured in the concept of being an "intrapreneur." This term appears to be first coined by Gilfford Pinchot and is elaborated in his book *Intrapreneuring: Why you don't have to leave the corporation to be an entrepreneur*.

Basically, an intrapreneur is someone who creates new stuff (food products in our case) without being constrained by the corporate culture and its standardized way of doing things. It certainly isn't the only way to climb the corporate ladder, but I do find that many corporate leaders ascribe to these tenets. Here is a list of "commandments" on how to be an intrapreneur. It is from Pinchot & Company's Web site (http://www.pinchot.com/MainPages/BooksArticles/InnovationIntraprenuring/TenCommandments.html.)

1. Build your team. Intrapreneuring is not a solo activity.
2. Share credit widely.
3. Ask for advice before you ask for resources.
4. Underpromise and overdeliver—publicity triggers the corporate immune system.
5. Do any job needed to make your dream work, regardless of your job description.
6. Remember it is easier to ask for forgiveness than for permission.
7. Keep the best interests of the company and its customers in mind, especially when you have to bend the rules or circumvent the bureaucracy.
8. Come to work each day willing to be fired.
9. Be true to your goals, but be realistic about how to achieve them.
10. Honor and educate your sponsors.

What ever your path up the corporate ladder, make certain that the trip is an enjoyable one and in line with your personal goals in life. I believe that enjoying your work everyday is much more important than where you end up on the corporate ladder.

I'll end with one last book recommendation. This is a very short book, so if you only read one, I suggest that you try this one. It is *The Radical Leap: A personal lesson in extreme leadership* by Steve Farber (Kaplan Publishing). Enjoy your career and make the world a little better place for your efforts.

References

Beveridge, W.I. (2004) *The Art of Scientific Investigation*. Blackburn Press, Caldwell, NJ.

Bradford, D.L. and A.R. Cohen (1997) *Managing for Excellence: The Guide to Developing High Performance in Contemporary Organizations*. Wiley, New York.

Farber, S. (2004) *The Radical Leap: A Personal Lesson in Extreme Leadership*. Kaplan Business, New York.

Goleman, D. (2005) *Emotional Intelligence: Why It Can Matter More than IQ*. Bantam Books, New York.

Pinchot, G. (1986) *Intrapreneuring: Why You Don't Have to Leave the Corporation to Become an Entrepreneur*. Harper Collins, New York.

Chapter 15
Employer Expectations: Could a Smaller Company Be for You?

Susan Hough

Good Things Can Come in Small Packages

You've probably heard a lot of stories about smaller companies such as the dictator, tyrant owner who micromanages everything you do, that small companies aren't as stable so you will have less job security or they can't afford to pay a decent salary. The information can be sometimes contradictory and confusing. One thing is certain, smaller companies dominate the landscape of job opportunities. It is estimated that two-thirds of all jobs are from small companies. Small companies are defined by the US Department of Labor as less than 500 employees (50 or less could better be defined as a startup company). Of course, a lot depends on the specific industry and how much automation a company may have. You will find that many in the industry will consider a small company to be under 200 employees, and a company that has 200–500 employees to be more of a midsize company. However you define it, there is a distinctly different culture and work environment between a small/midsize company and a larger national/ international company.

Are You Happier Being a Bigger Fish in a Little Sea

It is very easy to fall in love with the idea of working for the largest of the manufacturers: Kraft, Nestle, General Mills, Unilever.... These are names we all recognize, whose products we grew up buying. However, there are many more manufacturers we don't see. They dominate the smaller niches in the food chain, such as private labels, smaller brands, or the ingredients themselves that go into those famous brands.

S. Hough
The Masterson Co., Milwaukee, WI, USA

R.W. Hartel, C.P. Klawitter (eds.), *Careers in Food Science*,
DOI: 10.1007/978-0-387-77391-9_15, © Springer Science+Business Media, LLC 2008

Daily life in the smaller company can offer you some significant advantages over the corporate larger company. They can include the following:

- Since there are fewer employees, the smaller company typically will have a flatter organization, meaning fewer bosses to contend with and less of the associated bureaucracy. Less time is typically spent in meetings, and decisions are made more quickly. This means you have more time to focus on your work vs. navigating the landmines of corporate politics.
- You may find yourself working side by side with upper management or the owner themselves. You can learn from their example and management styles to help you understand those traits that allowed them to get where they are. Working side by side with upper management can lead to much more visibility and opportunities for recognition. Your ideas and suggestions can be given the attention needed for implementation.
- You get to see how the organization functions as a whole vs. seeing just one small corner of a division of a large company. The corporate and manufacturing may all be under the same roof, so you can see how customer service, scheduling, sales, purchasing, production, quality assurance, and research and development (R&D) all interact.
- With fewer employees, you will often have more job responsibilities and the ability to help out in areas outside your job description or department. You could find yourself on the line helping production, traveling to visit customers, or helping logistics on the year-end inventory taking. These experiences are a bonus in building a foundation for future employment and advancement.
- A growing small company's environment can be more exciting, dynamic, and fast paced. Smaller companies are often more nimble and versatile in order to compete in the marketplace. Changes and new product rollouts can occur more easily and quickly.
- It's an environment that forms strong bonds with your peers and fosters a real team environment. Often the small company has a less formal atmosphere.
- Don't fall for all the stereotypes that a small company can't compete with larger ones when it comes to salary or perks. There are always the cases where this may be true, but you really need to look at each opportunity on its own merits. In very rapid-growing small companies, you may, in fact, have more opportunity for career advancement or profit-sharing programs.

Welcome to Boot Camp. . .

With smaller, leaner companies, all employee contributions are critical to the success of the company, so with all the aforementioned positives can come some negatives. Visibility to the upper management can have the opposite affect if you are struggling to fit in or can't handle the challenges thrown your way. The smaller company can be more demanding, from the standpoint of getting their money's worth. They can't afford to carry extra employees on the payroll who can't deliver. You need to be honest with yourself in assessing whether the aforementioned environment excites you, or if you really are better suited for a workplace that is more predictable with more structure, defined rules, and responsibilities.

What You Bring to the Small Company

When reflecting on your abilities, look for the following traits that could help you predict your future success:

- Are you a self-motivated individual or do you need someone to tell you what to do each step of the way? Initiative and a "can do" attitude is a definite must.
- Are you a quick learner who picks things up quickly? The smaller companies may not have formalized training programs as the larger corporate giants. Your training may involve much more "on-the-job" learning. There is a certain expectation that you will adapt fast and speak up if you need assistance.
- You are good at multitasking and juggling several responsibilities or tasks at the same time. The more you assert yourself, show initiative, make suggestions, and show the desire to roll up your sleeves and dive in, the more successful you will be.
- You have good communication skills (written and verbal). You are a team player. You thrive on that feeling of camaraderie you get with working together with your peers.
- You are not afraid of a challenge, and can be flexible enough to handle changes in your work and responsibilities on a regular basis.

Getting that Job

There are a tremendous number of food companies out there. Don't wait for the ad in the newspaper or Internet posting to send in your resume. You can take advantage of the fact that a company has a job opening coming up soon or has delayed advertising for a job. An unsolicited resume will often get looked at much more closely than one that shows up with hundreds of others. If you have the right stuff, a company may take the path of least resistance and give you that chance first, to be what they are looking for vs. spending a lot of money on ads and recruiter fees. Companies have been known to see an outstanding candidate and even make an opening for them.

Before sending in that resume, do your homework first. You will want to consider the geographical area you will live. Is the company the right size for you? Are you open to all industries or do you want to narrow your field (meat, canneries, dairy, bakery, or confectionery). Does the company have a Web site you can visit? Many small companies (and some large companies too) may still be family owned. Try to avoid making assumptions or falling for stereotypes. Although there are the owners who are domineering and like to micromanage their employees, you can also find this in any company with the wrong boss. At the interview, try to get a feeling for how much freedom the employees are given to run the business.

A small company may be more flexible in the types of jobs they will consider you for. One area in which there is often an advantage to the food science candidate is in R&D/Product Development. The fact of the matter is that many of the larger companies may not even consider a candidate for this department unless they have a minimum of a Master's degree. Smaller companies are much more negotiable on this. Depending on your resume and work experience (including internships), they are more willing to entertain a bachelor's degree.

Stepping Stone to Greater Things

Don't forget that the smaller company can act as a great stepping stone to add to your resume. Today very few employees stay with the same company for a lifetime. If you plan your moves carefully, you can quickly gain a tremendous amount of experience and people skills to move up that ladder. Much can be said for the advantages of having interfaced with the upper management directly, having those diverse work experiences and broader responsibilities to build your career from.

Many people also move back to the small companies later in their careers because of the sense of control they can have over their work environment, not to mention the feeling that they can really make a difference at the company. A strong leader in a small company can positively influence it from a company that is always fighting fires, to one that is proactive and preventative. New ideas on policies and procedures to improve quality and reduce downtime, or increase sales, are more readily embraced. If you have a strong yearning to have a significant impact and change the culture of the company in a positive way, you may find that a small company is for you.

Chapter 16
Corporate Resources

Dennis Zak

Introduction

Learning is a lifelong journey. Despite the fact that you've just finished 4–8 years of college, your learning is just beginning anew with the first day of your new job. During the interview process, students invariably ask prospective employers about opportunities to continue learning in their new roles as employees. Students receive as many varied answers as they have interviews. New hires will soon understand that the main purpose of most food-related enterprises is to make a profit from the goods or services the company manufacture and/or sell. The enterprise is not in the business to specifically train or educate students. Education of students is left to the many high-quality colleges and universities across the country. Having said that, there are many opportunities to continue learning while being employed by the food industry.

It might be said that continuous learning is necessary to survive and grow in corporate America. Industrial organizations inherently believe that greater knowledge leads to more innovation and better solutions to problems. Tom Peters, author of *In Search of Excellence* and other business books, has stated that businesses and individuals must reinvent themselves every seven years to stay even with the competitive changes taking place in the market. Most industrial organizations support continued learning through a wide variety of formal and informal programs. Before discussing the obvious formalized learning programs, a discussion of the learning that should take place daily within standard business processes is warranted.

D. Zak
TMResource LLC, Doylestown, PA, USA

R.W. Hartel, C.P. Klawitter (eds.), *Careers in Food Science*,
DOI: 10.1007/978-0-387-77391-9_16,

Corporate Culture: Learning Through Experience

By nature, humans will develop processes, formally or informally, for anything they do more than once. These processes and their continuous improvement are responsible for the growth of human society since the beginning of mankind. In addition to processes, all groups of humans develop relationships with other members of the group. Some of these relationships are strong and trusting, others are weak and filled with anxiety. Humans also have relationships with themselves, which psychologists call emotion. Just as we measure intelligence with the Intelligence Quotient (IQ) tests, psychologists measure emotion and emotional stability with an Emotional Quotient (EQ) test. Displaying appropriate emotions in business situations is extremely important in developing trusting relationships and working in teams.

Humans and organizations also have resources. Employees, financial assets, computers and other equipment, good will, and brands are all organizational resources. These three big boxes—processes, resources, and relationships—are put together in what can be called a human endeavor model. All human endeavors start with some input, some human longing, some need. Humans then use resources, processes, and relationships (team work) to produce an output. However, output is not outcome. Competitors and many other environmental events or activities come into play to influence the outcome. As stated above, an individual is a resource, tools are resources, and things within the environment are resources. All this is combined in what I refer to as the "human endeavor model" shown in Fig. 16.1.

Test the model. Most anthropologists indicate that early man was a hunter–gatherer spending most of his time trying to find enough food. The input for this early man was hunger. He searches his environment for resources and sees a rabbit. He also finds a stone and throws it at the rabbit (output). He

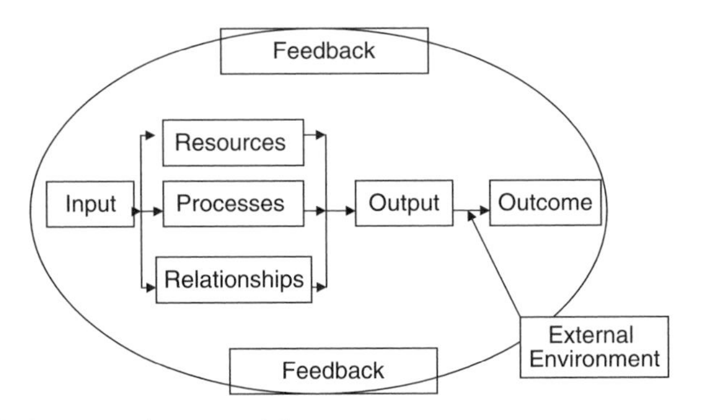

Fig. 16.1 The human endeavor model

misses, throwing the stone behind the running rabbit (outcome). Emotionally (emotion or ego is one's relationship with oneself), he is upset and still hungry. He uses the feedback loop and modifies the process to throw the stone in front of the running rabbit (output again), hits the rabbit, and is thus successful in obtaining food (outcome) to satisfy his hunger. This is a very simple application of the model. Stronger interpersonal relationships (teamwork) developed when man learned to drive the prey toward the hunters who used spears, bows, and arrows derived form the resources around them. There are a multitude of processes and subprocesses, relationships, and resources in any organization. This is why we have divisions, departments, groups, and teams to better focus organizational endeavors. It is the totality of how we do things and treat people that make up an organizational culture. All groups of people including married couples, families, schools, and clubs develop organizational cultures. Habits are what individuals do the vast majority of the time. Cultures are what groups or organizations do the vast majority of the time. Circumstances can alter cultures from time to time, but most groups fall back on the tried and true processes for achievement. If you, as a new employee, do not understand the culture, it will be very difficult to have a successful career in the organization.

Successful food businesses, particularly the larger ones who have been in business for a long time, have highly developed cultures or processes for what they do. Smaller entrepreneurial organizations or companies have less formalized processes, typically derived from the owner or entrepreneur. In most companies, there are no formal, structured classes to learn these processes. Indeed, some organizations do not realize these processes actually exist. They engage in them because "this is the way we do things around here." The new employee will learn the processes through relationships with others in the work group. Longer term employees will show new hires the proper forms to fill out, which meetings to go to, and which activities are most important. This is the socialization that takes place as the new hire strives to belong to the work group. The new hire must learn these processes through mostly informal means to be successful in the organization.

New hires should be proactive by searching out and understanding the preferred processes for communication, feedback on performance, reward and recognition, promotion, assignment allocation, team building, and learning through formal, informal, and experiential situations. Of all the processes or systems in any organization, communication, and recognition and reward systems are the most important to advancing one's career. Some companies have very formal written communication processes where others are very verbal and informal. Each new hire must quickly learn the differences between what is espoused and what really happens. Look for the patterns in how documents are written and how presentations are made. New hires need to quickly learn the formats and amount of detail preferred by the various levels within the organization. Often there are formal business communication meetings but employees learn everything that is presented at the water cooler. New hires should pattern their communications from those who were recently promoted. Their behaviors and skills are obviously accepted by the organization.

Most organizations have some type of formalized employee review process on an annual or biannual basis. In some organizations, this process is paramount to a successful career. In other organizations, it is simply an insurance policy against unfair labor practice lawsuits. Career advancement potential is determined by other criteria. In some companies, employees get formal, written biannual performance reviews that give little constructive feedback. The real valuable career feedback is derived from the frequent informal conversations between the supervisor and the subordinate. One of the most important things to understand about any organization is how it measures success: what the organization considers important. Employees want their work to be useful and valued by the organization. If the new hire does not understand what the boss and the organization considers important, it will be difficult to advance one's career. Almost universally, corporations want profits.

Profit Is not a Dirty Work: Understanding Accounting

Value Creation

There are two parts to competitive economics: costs and value (Fig. 16.2). As a new hire, you are a cost. You have a salary, benefits, a work space, computer, cell phone, an expense account, etc. The organization expects you to return value to the organization in excess of the overall costs. Profits come out of the value created by the organization. The company's competitive position in the marketplace determines the value proposition it offers the customers and the consumers. If your organization offers a product or service with a better value proposition to the customer and the consumer than your competitors, your product or service is purchased. Consumers who perceive value in the purchase will repurchase the product frequently. Product

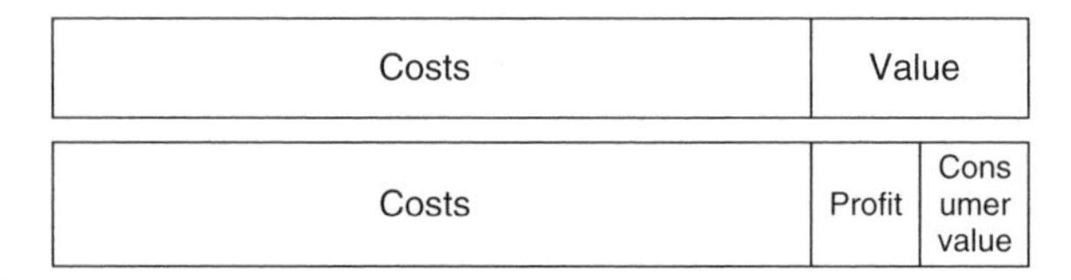

Fig. 16.2 Competitive value proposition

quality is inherent in the consumers' perceived value. Products that have a low perceived quality, generally have lower value (lower price) in the eyes of the consumer. However, product quality is much more than *fitness for use*. Quality is meeting or exceeding customer or consumer expectations. Think of the value you would obtain if you purchased a Cadillac at the cost of a Chevrolet. Both are fit for transportation. On the other hand, imagine the dissatisfaction in paying Cadillac prices and receiving a Chevrolet. This does not mean that businesses making these products are not successful. McDonalds may not make the best tasting hamburger relative to, for example, *CheeBurger, CheeBurger*. The purchase price ($4.50 vs. $7 for *CheeBurger*) and other aspects of the McDonalds business model (speed of service, kid friendly, cleanliness, etc.) meet consumer value expectations and thus McDonalds has a very successful business, generating profits, and increasing shareholder equity.

Accounting

In very simple terms, accounting systems look at the financial health of the organization. It subtracts the costs from revenue collected and determines if any profit results. Note that there may be more than one set of books. There are accounting ledgers for IRS and for investors. These ledgers are governed by good accounting practices. They should be very similar from organization to organization. There may, however, be a different set of books that the organization uses to *manage the business*. Some organizations are looking for a return on investment (ROI), some are looking for profit as a percent of sales, and some are looking for a return on total assets (ROTA). Assets can be based upon current value or historic value (purchase price), and historic value can be may be based on a tax depreciated value. The accounting system used to manage the business may look very different from organization to organization. It is very important to determine where and how you impact the accounting system.

In general, most accounting systems look somewhat like the following:

$$\text{Sales} - \text{Costs} = \text{Profit before Taxes} - \text{Taxes} = \text{Profit after Taxes}$$

Revenue is obtained from the sale of a product or service, and costs are subtracted from that revenue. Profit or loss before taxes is the result. Taxes are paid and you have profit after tax. Business organizations cannot survive long without profits. Profits provide funds for new investments to remain competitive. Profits are a source of taxes for communities. Profits provide returns to investors. Profits ensure long-term salaries and benefits for employees and payments for vendors who supply ingredients and services to the organization. Even nonprofit organizations like Newman's Own want profits from the sale of their products so these revenues can go to charitable causes. Remember that profits come from the competitive perceived value in a product or service as determined by the end user.

Costs include all wages, salaries, benefits, raw materials, advertising, promotions, transportation, utilities, interest, energy, etc. There are hundreds or perhaps thousands of entries into a typical business accounting system. There is no magic on where to account for all these costs. The accounting system is generally parallel to the organizational structure to hold the various managers accountable for costs in their areas, i.e. advertising and promotional costs are in a marketing cost center, and line trial costs are in a R&D cost center. Managerial accounting systems are as variable as the organizations that use them.

It is easy to see where you, as a new hire, impact the cost side of the ledger. Your salary and benefits fall within a specific functional area or cost center. Depending on the job, it may be more or less difficult to see how you impact the revenue or value part of the accounting system. If your job is technical sales, you directly impact sales revenue. Your technical input may also impact the quality and value of advertising through more targeted print and media advertising materials. You may even reduce returns through advising customers in better use, handling, and storage of your products.

If your job is R&D, you can introduce new products with a direct addition to the revenue line. Research and development employees can also reduce raw material costs through finding functional but less costly materials. You can reduce direct labor or raw material overusage by installing more efficient processes. If you are in quality assurance, you can reduce raw material overusage costs by providing more effective operating practices for operators, by providing storage and handling practices that extend product shelf life and consumer value, or by providing quality systems that reduce returns. Costs saved go directly to the bottom line. If a company is making 5% gross profit, it will take $20 in sales to be equivalent to $1 of cost saved.

Employees can impact several accounting areas within a given fiscal year dependent on the number and type of projects they work on. Knowing where and how much you impact the profit picture of an organization will keep you engaged with the organization, give you job satisfaction, and demonstrate the value you have created for the organization. The boss is your first resource in aiding you in understanding the accounting system in the new organization. Have your boss recommend someone in accounting to further understand the system. The new hire should understand how the budgeting is done for the work group, the department, and the organization as a whole. Likewise, you need to learn the specific terms used and what they mean. The specifics of gross margin, prime margin, ROTA, and others will be different in different organizations.

Connecting to Business Goals

Strategic Plan

Many organizations have a strategic plan covering the next five years. In broad terms, this plan covers the growth projections, resources, and specific directions the organization is taking to achieve its long-term vision or 20–30-year goals. Good plans will have specific project activities with associated goals for acquisition, growth, productivity, and cost containment. The plans are typically updated yearly as the competitive environment changes and goals and resources have to be adjusted accordingly. Plans that are radically adjusted on an annual basis should not be considered "strategic" plans. The radical adjustments indicate the organization is not charting its own course, but simply reacting to the changing environment. Organizations are a collection of individuals, and as with individuals, the best indicator of future behavior is past behavior. Organizations that consistently miss annual goals, but keep the plan will continue to miss plan goals and inevitably lose market share and struggle.

A good way to connect to the organization is to connect with the strategic plan. If there is a strategic plan, you should read a copy and ask both your boss and associates in your work group how well the organization has done in the past in meeting plan goals. The work group or department should have a section in the strategic plan. This section should illustrate the type of activities you are to do in support of the plan. You should look for ways you and your work group can specifically and directly support the plan goals.

Your annual objectives should flow from the "Objectives" section of the strategic plan.

Objectives

Once you understand the strategic plan and your work group's part in support of it, you can put together your own specific objectives. Your objectives are what you specifically are doing to add value to the organization. They are not team goals or team objectives (we will talk about teams and team work later). Your objectives are the activities you are responsible for in support of the teams in which you participate. Your objectives are the activities on which your performance is measured: your annual renewable contract with the organization. A written objective is essential for performance assessment. Not only is it a written contract between the manager and subordinate, it is a boundary document that allow both the manager and subordinate to focus. This makes the often-tense performance discussions easier for both the manager and the subordinate. Moreover, they are a planning and time management tool for the associate.

There are many formats for developing objectives. The following is a discussion of three formats: "my" personal favorite, which has no specific name; SMART; and Covey. The format I used was derived over 20 years of structuring objectives. It is summarized as follows:

Title
Objective with measurement criteria
Business benefit
Approach
Degree of difficulty: Low, Moderate, High
Resources
Milestones

Title

The title is for administrative purposes. Managers like project titles to better organize the portfolio of objectives for their subordinates. Title is not necessary for the subordinate.

Objective

Your objectives should be specific to your role in the organization. Let's say a team is charged with launching a new product called XKX. Marketing typically defines the concept or words that describe the product. Financial associates usually determine cost parameters for success within the business. For the product developer on the team, an objective might read: Have developed a product recipe and process for project XKX achieving a 5.8 hedonic score (how well consumers like the product), a 70% product concept match and having a 1-year shelf life with a raw material and packaging cost of 55% of the selling cost within 20 months. Note that the objective is written in the past tense. Rather than "to develop," it reads "have developed," indicating it is in the past. This, when coupled with milestones makes the objective more definite. Note also that the measurements are included with the objective. Measurements need to be realistic. How hard would a runner work if success only came with a three-minute mile? How hard would you work if to achieve a business goal of all the attributes of whipped cream with zero calories? Measurements need to be realistically achievable. Getting a hedonic score of 6.95 out of possible 7.0 is not realistic. Getting a hedonic score of "at or greater than" the category average is achievable. If the category average hedonic score is 5.6, a target hedonic score of 5.8 is realistic. Getting raw material and packaging costs as well as shelf consistent with current business norms is realistic. This does not mean that you should not take on challenging projects. Many once thought space travel unachievable. It did however take a long time. When measurements are this specific, the beginning of strategy or plan for achieving the objective starts to form (see Approach section).

Business Benefit

Before a discussion of approach is undertaken, the benefits to the business need to be determined. This is where the new employee connects to the company accounting system. Look at where you impact the cost side of the ledger and how your output leads to value for the organization. In general, business organizations do things for four reasons: (1) to obtain fundamental knowledge, (pure or academic research), (2) grow the business, (3) make the business more efficient or productive, and (4) to generate good will. All of these have the potential to improve the bottom line and add value. Fundamental knowledge leads to patents and innovative new products. It is easy to see how new products and efficiencies lead to added value: they increase revenue or reduce costs. Goodwill is being a responsible corporate citizen. It is taking part in industrial and community organizations. Goodwill is helping train smaller companies relative to new regulatory standards. Product recalls hurt the entire industry and not just the company whose products are in question. Trade groups lobby for or against legislation based on how they feel the legislation will impact the whole industry. Goodwill is like an insurance policy, helping the organization avoid future costs.

Approach

The approach or strategy section of the document discusses the general course of action to be taken to achieve the objective. Some actions or activities are already defined by the required measurements noted in the objective. Others are defined by the cultural processes of the organization. Others may need to be filled in, for discussion with and approval of the supervisor. For the new product XKX, there is most likely some initial bench-top development with reviews by marketing and other key internal team members or perhaps some early qualitative consumer tests. The process needs to be developed and tested. Line trials have to be conducted. Remember the organizational processes discussed above. This is where they are implemented.

Degree of Difficulty (DOD)

Not all projects are created equally. Some are more challenging than others. Projects that require new science or new process discoveries are certainly more challenging than those that only require implementation of an existing process: i.e. a flavor line extension of an existing product. The former would be a high degree of difficulty (high DOD) and the latter a low degree (low DOD). Likewise, implementation of a new companywide "lean manufacturing system" will be more challenging than removing a single bottleneck in one production line. Many others lie in between. A well-balanced portfolio of projects should have a balanced DOD for each employee. Employees higher up the organizational chart should have projects with higher challenges. These employees cost more and are challenged to add more value to the organization.

Resources

After the approach and the degree of difficulty have been determined, resources can be estimated. They should include the individuals time required for this project and any resources not within the individuals control. If consumer test funds are in another budget area, agreement needs to be made ahead of time that a portion of those dollars are available for this project. If other departments are involved, agreements are required ahead of time that their resources will support this project. Most conflicts in organizations come from lack of agreement on what is to be accomplished and who is responsible and who pays for what. These objective structures are designed to prevent that conflict between subordinates and between the various work groups in an organization.

Milestones

Lastly milestones are listed. These are the deadlines for various key activities defined in the approach section. Most companies require quarterly reports to be consistent with financial reporting requirements. Objective milestone dates should be consistent with these reporting requirements. This is another connection to the accounting system. Be advised that it takes time to roll up activity reports from the bottom to the top of an organization. If financial reports are required in April, July, October, and January each year, your reports may be required in February, May, August, and November.

In general, the higher up you go in an organization the longer it may take to complete an objective. Developing a new analytical method may take a bench chemist 6 months, whereas changing an organizational culture could take an R&D vice president several years. Putting a new product on the market in 2 months may be unrealistic for a large company that needs large pieces of equipment, but very realistic for a small company who only needs to go to the restaurant supply store and buy several new pots.

Table 16.1 SMART objective format

S	Objectives – Should specify what you want to achieve
M	Measurable – You should be able to measure whether you are meeting the objectives or not
A	Achievable – Are the objectives you set, achievable and attainable by you?
R	Realistic – Can you realistically achieve the objectives with the resources you have?
T	Time – Deadlines/dates for completion

Structuring objectives should be a deliberate, thoughtful process. It should not, however, result in an overly detailed 10-page document. Each formatted objective should take no more than three-quarters of a page.

The **SMART** acronym and Covey format are summarized in Tables 16.1 and 16.2.

Note the similarities of the formats. All specifically define the end point. All specify measurements and deadlines. Accountability in Covey equals Measurement in SMART. The Covey model has a consequence section, SMART has a section on realism, and my favorite makes a stronger connection to the business and has approach or strategy section. All are very good and useful models. Even if your business or boss does not require written objectives, structure them yourself. In the process of structuring your objectives, you will be better connected to the business, better structure your time toward what is important to the business, and more effectively demonstrate your value to the business.

Regardless of the model, you will find that a significant amount of time is required from the boss and the subordinate to structure the objectives. Once in place, the subordinate is free to pursue the strategy to complete the objective. Most bosses are content if subordinates are on time, within budgets, and achieving as planned. Bosses want to know if you are ahead or behind schedule. If ahead, resources may be available to aid those projects that are behind. If behind, particularly in a project key to the business plans, bosses will want to change strategy or secure more resources for the key

Table 16.2 Covey objectives format

Desired results	Clarify the end objectives, goals
Guidelines	Specific boundaries, deadlines for completion of activities
Resources	Human, financial, technical resources available to complete objectives
Accountability	Standards and methods of measurement for accomplishment
Consequences	What happens with achievement/nonachievement

project. Bosses do not like surprises. Keep them regularly apprised of your status on your objectives.

Bosses/Supervisors/Managers

Other than yourself, bosses have the most significant influence on your career. You are and will continue to be the most important person in your career. You always have the ultimate decision power, the power to leave for another position. But, bosses also have power. Regardless of how much you perceive they know, or how much they help you, bosses have power associated with their position. If they are effective in their jobs, consistently meeting or exceeding work group goals, and their subordinates readily get promotions, they have legitimate power that comes from skills and knowledge. Bosses make decisions on how the work is allocated within the work group. They typically control the distribution of resources and assignments. They encourage or discourage training assignments or activities. Bosses are the ones who ultimately write your performance appraisal, the ones who assess your value to the organization. In a 30–40–year career, you will run across supervisors who are dictatorial, authoritarian, and control everything and ones who are very removed, offering no direction what so ever. I have had bosses who thought their concepts and decisions where guided by God and others who challenged nothing and input nothing. Imagine the conflict that arises with a young scientist trying to challenge God with scientific logic.

The vast majority of bosses will lie between the extremes. Learn from their strengths and weaknesses. The authoritarian boss mentioned above was an extremely hard worker. He anticipated challenges to his concepts or positions and was ready with answers to meet those challenges. He was disciplined in managing the activities of the group. He had to be to maintain control. The initial structure of my favorite objective setting format was derived from him. This boss had great planning, organization, and influence skills. On the other side of the ledger, there was no room for the subordinates input, little professional satisfaction, and little or no directed growth. Information was held close by the boss. Subordinates had difficulty connecting to the organization. He was very poor at developing others, delivering creativity and innovation (only one idea mattered) and dealing with conflict within the work group. Managers in most organizations are charged with growing subordinates. Organizations want employees to be able to contribute more, to add more value. That increases the value side of the cost value proposition available to the customer and consumer and the business. Look for patterns of behaviors from the boss. Learn to discern the positives and negative skills of your boss and other bosses associated with the teams you work on. Most college graduates will some day be a supervisor of others. Admire and embrace those positive skills. Recognize and tolerate the negatives remembering not to employ them when it's your turn to supervise others.

Teams

There are still some jobs in industrial organizations for individual, self-directed research. This type of research is very similar to what is currently undertaken in graduate programs at universities. However, the vast majority of the work in industrial organizations takes place via cross-functional teams. Team work is essential to achieving goals in industry. Teams form to forward specific projects, are disbanded if the project is achieved or canceled and form again to support the next project. Team membership may change slightly or radically depending on the new objective. Some teams stay together indefinitely with one or two members rotating in and out annually. Teams also offer also a great opportunity to learn about the roles of other functions as well as the processes they employ to achieve their goals. Teams are where new employees will learn about the culture and processes of the organization as a whole. New employees can then see how the puzzle pieces of the entire organization all fit together to obtain the desired value picture.

There are several different models that discuss team building. All cover similar aspects of building relationships. The Tuckman model goes back to 1965, but it is simple and easy to remember. It has four stages as described as follows:

Stage 1. Forming. When teams first get together, members are generally cautious and uncertain. People do not know or trust each other. Everyone tries their best to look ahead to the end objective. The leader must set the team focus.

Stage 2. Storming. Inevitably the process begins to heat up under the pressures of work and conflicting perspectives. Members jockey for influence. With no prior experience, trust is lacking. Members challenge each other concerning roles, allocation of resources, goals, budgets, and deadlines.

Stage 3. Norming. As activities progress, people get to know each other and see others meet goals. Trust grows. Team members reconcile and agree on norms or processes like decision-making processes, resource allocation, timing, and quality standards. A "norm" is something everyone understands. Norms are the formal and informal rules that make up the operating system of productive work. These are the work processes we discussed earlier.

Stage 4. Performing. The final stages of team development involve using all the experience and understanding with each other to get results for each other and the organization.

High-performing teams can easily absorb 30–40% turnover annually. The remaining team members rapidly "socialize" with new team members. New members normally want to belong and initially do not to want to change team norms. New members generally wait until they are accepted by the team to propose significant change. New team members must earn their stripes. They must prove their value to the team. If turnover exceeds the above values, the team building process begins anew. Trust and the associated team processes must be rebuilt. The trust among the team members lasts long after the team has completed its task and has disbanded. Teams can reform rapidly even after several years if 60–70% of the members have worked with each other on previous projects. The trust that develops extends beyond the current organization even if one member leaves. Often one employee attaining a new position in another organization brings other trusted associates with him to the new organization.

This first section has covered learning from experience. The learning rate is greatest at the beginning of each new experience or new job. Obviously, the broader the job experiences the greater the learning. More varied jobs, with different challenges means more learning. Learning will be greater if employees take risks and get out of their comfort zone. However, if learning does not keep pace with the new job requirements, performance will suffer and the value added will soon be less than the cost.

Corporate Competencies

Over the past several years, organization or business competencies have gained wide acceptance in the US industry. Business competencies are defined as the skills and capabilities that allow the company to attain its current position in the market place. A company may have excellent manufacturing capabilities. It may be the least cost producer of a particular product. It may have an outstanding customer service resulting in 100% order fill and on time delivery to its customers. A company may have sales and/or marketing capabilities making its products top of mind in the eyes of consumers. It may have distribution capabilities, meaning its products are found everywhere or a company may have innovation competencies making it a market leader in new products. It may have all or some combination of the aforementioned capabilities or others not listed. In the consumer electronics industry, it has long been said that Sony is a very innovative consumer electronics company. Panasonic is a fast follower, lower cost producer of similar consumer goods. These are both corporate competencies. Both companies are very successful in different ways, with different competencies. They are structured differently and have different business processes or competencies.

Corporate competencies are underpinned by the competencies of the individuals in the organization. It is the collective competencies of the employees that result in corporate competencies. Understanding competencies provides an organization with a framework to maintain its current market position and also to change consistent with its vision. If Panasonic wants to become the electronics innovation leader, it will require different individual competencies than it currently has. Individual competencies are the capabilities, abilities skills, knowledge, and expertise that allow an individual to be successful in their role in an organization. There are hundreds of competencies. They are usually divided into two groups, technical and nontechnical. Technical competencies are the skills one learns in college, in food science, engineering, finance, or graphic arts. The technical skills are best maintained through technical conferences, specific technical seminars, and relationships with universities.

The nontechnical skills are what were once called soft skills: leadership, business acumen, time management, directing others, project management, ethics and values, creativity, innovation management, etc. The book *For Your Improvement* by Michael Lombardo and Robert Eichinger is a good source for competency definitions as well as measures to determine if individuals have the skills. Technical skills, particularly for the new graduate, are the cost of entry into your initial position in an organization. These are more readily seen in the accomplishments of the new hire, the items listed on the resume. Managerial or nontechnical skills are more difficult to assess but are needed in most cases for new hires to advance in an organization.

Corporate Universities

To maintain or improve existing competencies or change to meet the challenges of a changing marketplace, many businesses have developed the concept of corporate universities. The corporate university allows organizations to formalize learning about corporate cultures and more rapidly integrate the new hires. It is also a more rapid way of changing cultures. These corporate universities offer a catalog of formalized courses supporting both technical or functional competencies and also nontechnical competencies. Each corporation has its own culture, its own way of doing things, or its own processes. It is an efficient way of changing the corporate culture through the addition of new courses.

The courses may be fully developed and implemented by the organization, contracted to outside corporate education firms, or provided in association with local universities or colleges. Course formats range from typical classroom settings of several hours to several days, to e-learning venues offered at whatever pace the student desires. In the food area, many have laboratory exercises where the employee students learn to make their own or competitors' products. Manufacturing personnel learn how the financial group keeps track of the financial health of the organization. Marketing and sales associates learn how products are made. R&D associates learn how purchasing works to keep costs under control. Each functional group trains the others in their work processes. Equipped with this knowledge, cross-functional teams form more rapidly, resulting in more effective implementation of business projects.

The corporate university may have a relationship with a local university where employees can obtain an advanced degree through typical tuition reimbursement programs. Students can go to the university to attend the class or if interest is high, the university professor can come to a business conference room to deliver the class. In some cases the class is delivered electronically. Many universities offer week-long executive programs uniquely designed for a specific company. Moreover, many university professors are invited to give short, day-long seminars or teach classes on a periodic basis over a year. New employees should be prepared to take advantage of these learning opportunities as growth in an organization depends on demonstrating both technical and managerial skills. Everyone should have learning objectives supporting their personal growth as well as business objectives. A good rule of thumb is to attend at least one technical and one managerial learning course every year. Because technical learning is more closely related to the existing job and current business projects, these opportunities are attended more frequently. However, except for the few that have fundamental research positions, demonstrating managerial competencies may, in the end, be more influential to your career.

The class room training courses offered by the corporate university are only the beginning of attaining the needed skills. Classroom training is only a small portion of the learning process. Experience is the most significant part of learning nontechnical skills. Lombardo and Eichinger in their book *The Leadership Machine* indicate that 70–90% of the learning takes place on the job. When the strategy or approach part of the learning objective is defined, classroom training should be followed by an "on the job experience." For a new hire, with no one reporting directly to him or her, a class on directing others can be followed with an opportunity to manage an intern to gain the experience. A class on creativity, followed by cofacilitating creativity sessions on actual business projects, is a good strategy to learn and demonstrate creativity skills. What is learned in the classroom is valuable, but what is experienced is far more valuable. Imagine class room training on learning how to ride a bicycle. Instructors will tell you to turn the wheel in the direction you are leaning or falling. They can tell you how it feels to be leaning or falling. But, it does not mean much until you actually feel the sensation of falling. Once learned, the skill is retained for a lifetime. The same is true of experiences with nontechnical skills.

Corporate universities typically do not offer leading-edge technical training. Because this is so job or individual specific, it is more effective to have employees attend scientific seminars or symposiums. If the position requires maintaining leading-edge scientific skills, it will be the employee's responsibility to find and attend the appropriate seminar. Relationships with institutions engaged in leading-edge research are typically encouraged. Many businesses readily support seminar attendance and leading edge research financially through either trade organizations or directly with researchers. Businesses are, however, looking for value to be added to the organization for the investment. Value addition does not have to be immediate. Over the long haul, businesses want to see employees being able to contribute more value. Individual employees are responsible for attaining and demonstrating the skills to do so.

External Organizations

An excellent source of learning comes from organizations external to your company. As stated above, every group will develop a culture. The corporate culture can become self-limiting, doing things the same way for long periods of time. By default, successful companies are successful by doing things the way they do them. It is their core competency. However, competitors may have found new efficient processes or developed new more effective competencies. Many organizations only change significantly when some external circumstance or competitive threat causes them to do so. Automakers in Detroit are forced to change today due to foreign competition and high labor and benefit costs, which resulted in poor value and profits.

A great way to prevent this forced change of culture is to bring in new process and/or technologies into a company through participation in external organizations. There are numerous technical organizations (Institute of Food Technology, American Dairy Science Association, American Association for the Advancement of Science, American Institute of Chemical Engineers, Association of Analytical Chemists, etc.). There are trade organizations (National Food Processors Association, National Confectioners Association, National Restaurant Association, American Association of Meat Processors, etc). There are management associations (American Management Association, Academy of Management, Academy of International Business, etc). In the management area, there are organizations that target specific functions in organizations. The American Society of Quality targets quality improvement; the Industrial Research Institute explores ways to improve R&D management; the Society for Human Resource Management strives to improve human resource management in organizations. The International Food Informational Council is an organization whose purpose is to communicate science-based information on health, nutrition, and food safety for the public good. There are external organizations that cover *all* aspects of business.

Most, if not all, of these organizations have publications, Web sites, annual meetings, and share groups where members share information. Many have local chapters with dinner meetings with guest speakers discussing topics of interest to the group. These meetings are great places to network with employees of other companies and learn generally what is transpiring in your specific area of responsibility. This networking will become the major vehicle for job changes as your career advances or you are caught in company reengineering or down sizing. To get the most out of these opportunities, make a concerted effort to meet new people by choosing a dinner table where you know few people.

Most companies encourage your participation in these groups through funding dues, publications, and expenses to attend seminars and trade shows sponsored by the external groups. Many, but not all, companies will encourage employees to take leadership roles in these organizations acting as officers or committee members. This gives the employee an opportunity to demonstrate leadership or soft skills to both the employer and to other companies as well as greatly extending the individual's network. You must remember, however, that companies strive to provide value to customers and consumers through the products and/or services they provide. The company is looking for you to provide value back to the organization through bringing in useful information about new ingredients, technology, or business processes. Provide more value to the organization by writing up a brief report on what you learned. Writing the report helps you remember the new information, as well as giving others access to it. Most employees will have objectives and priorities that are more important to their success than participating in external organizations. Spending excessive time with these eternal organizations will have negative consequences on career growth. You must find a balance in completing current business objectives and projects, staying current in your current job responsibilities and developing and demonstrating skills for the next position on your career path. Including participation in external organizations in the formal objectives document described above can help in finding that balance.

Summary

Remember that you are the most important person in your career. Others have significant influence on your career, but you are in control. Learning is a lifelong journey. You should set learning goals. Be proactive in achieving those goals. Organize your learning objectives the same way you would organize any other objective. Step out of your comfort zone and search for new assignments and job opportunities within your organization. Try out new processes to determine if efficiencies result. Build new relationships and networks to help you now and in the future. The experience you gain from many different jobs and the trusting relationships you build will greatly increase the value you add to your current or future organization.

References

Covey, Stephen R., 1989, *The Seven Habits of Highly Effective People*, Free Press, Division of Simon & Schuster, New York, London, Toronto, London.

Lombardo, Michael M., and Eichinger, Robert W., 2000, *For Your Improvement*, Lominger Inc., Minneapolis, MN.

Lombardo, Michael M., and Eichinger, Robert W., 2001, *The Leadership Machine*, Lominger Inc., Minneapolis, MN.

Tuckman, B., 1965 Developmental sequence in small groups. *Psychological Bulletin*, 63(6), 384–399.

Chapter 17
Making Your Way in a Company

Christine Nowakowski

You've worked hard in school and now you find yourself starting your career in industry. The beautiful thing about food science is that you can come from a variety of disciplines with one common thread, an interest in food. Unfortunately, that is precisely the challenge I have trying to describe a typical job in the food industry to you. Food science is a catch-all for engineering, biotechnology, chemistry, material science, microbiology, and law among other things. This diversity allows food scientists to fit in a company in a variety of positions within quality and regulatory and research and development (R&D). No matter where you work, government or industry, large or small company, you will require interpersonal skills and technical excellence in order to succeed. In general, the demands of a full-time job require skills that are not taught in college and how well you handle the demands of the job strongly influence one's personal sense of accomplishment and life balance. It is critical to build in external input to maintain perspective and clearly see the next steps in your career. Here are some things to consider as you start out in your career.

Work Skills

Level of education, personality, and market demand for your particular skill set will, in some way, influence the number and types of employment offers you receive. Technical excellence in your specialty is critical. For example, in food chemistry, it is expected you know details about Maillard browning, Arrhenius kinetics, water activity, and viscosity, to name a few. As a food microbiologist, you will need to demonstrate skill in understanding growth

C. Nowakowski
General Mills, Minneapolis, MN, USA

R.W. Hartel, C.P. Klawitter (eds.), *Careers in Food Science*,
DOI: 10.1007/978-0-387-77391-9_17, © Springer Science+Business Media, LLC 2008

rate, food safety, water activity, and Arrhenius kinetics, among other related topics. Knowing your skill set well will help you solve many issues, whatever your job function. It is also important to know how to learn what you need about a topic you're not familiar with. That is, you need to learn how to find information and sort through what you find to figure out what is important.

One key difference about post-graduation from being a student is that you don't need to have the answer immediately. It is quite okay to say, "I don't know" when asked a question. Practice this, really. A skilled response to a question that you aren't sure of the answer is, "I don't know, but I know you can look at this reference or talk to this other person." This response shows that you can help others and be a positive force to the team in ways other than your specific technical excellence. It also shows integrity and builds trust with others at various levels (supervisors, coworkers, direct reports).

Another quality that is helpful is leading innovation. This can come about by communicating a compelling vision. In order to do this effectively, you need to have an understanding of how your project fits in the larger corporate picture. From there, successful developers create a timeline or detailed game plan to communicate how they are going to accomplish their goals. There are software programs to help you develop timelines and these can be useful if you prefer that style of organizing or if you have many complex tasks to accomplish. Or, you could simply make a list of things you need to do, attach due dates to them and a list of who needs to know about progress and the timing of those updates. One thing a new person has to keep in mind is that many others will be involved in your work and this is quite different from being a student, where many of your tasks are independent.

Finally, a drive to be successful is important. This ideally creates an enthusiasm for progress that attracts positive effort from others. The majority of your efforts in a corporation involve others. A successful person will be able to connect and leverage others' strengths to attain the project goal.

After assessing your skill set, it is helpful to have in mind some critical criteria as you choose your employment. This is important in that you need to know how the present situation fits with your overall life and career strategy. Getting that first job is great, but a good fit is even better. If the present employment will not be a comfortable fit, then it is critical to start planning your next career step before your employment environment impacts your overall view. You will be spending a lot of time at work and it will influence your life outside of work. A good fit can be a positive force whereas a poor employment fit can negatively impact life. I had developed a questionnaire that I applied to all companies I interviewed with as a means to fairly compare each. One question I constructed was,

"What is your paternity leave policy?" Every company I interviewed with used the latest business buzzwords: Work–life balance. And, many showed examples of just that. But my question had a bit more to it, I thought. It was a great test to see if they were actively listening to me. I'm a female, why would I be asking about paternity leave? I got many answers about maternity leave and then I'd redirect them to my question. Some companies had a policy in place, some were reviewing such a policy, and one company would not consider this policy. I'm not saying that there is a right or wrong answer. I just wanted to know what their perspective was on this (at the time) new concept in work–life balance. Additionally, I've written down my short-term career goals (2 years), long-term goals (5+ years), and legacy theme (what I want to say about myself after 25 years in the business). By revisiting these biannually, I can measure how good a fit my current employment is and how well this fits into my overall goals. I have seen others who have not taken this multitimed view find great short-term employment opportunities that did not lead them to a sustainable, satisfying career.

Feedback

There are five common ways to get external feedback: informal/impromptu feedback, periodic reviews, network checks, 360° reviews, and formal performance appraisals. Informal/impromptu feedback is best described as simply asking people you are in dialog with about how you're doing. Be specific as to the type of feedback you'd like to get. At the end of a meeting, for example, simply ask: How did you think that meeting went? Is there anything I should consider changing or anything that went particularly well? Jot notes to yourself and consider the feedback. I like to save these notes in a file to reflect upon to identify trends.

Regular periodic reviews with supervisors are critical to your success. If you have a supervisor who is crunched for time or simply doesn't see the value in regular meetings, take charge and organize them yourself. Biweekly or monthly meetings are usually sufficient. My goal for these meetings is to inform my supervisor of what I've been doing, ask him or her what's the strategy in the near and not so near future and how I'm synching up with expectations and stated goals. Create the agenda or jot notes to yourself prior to an update meeting. This allows you to reflect on two weeks worth of work and supports self-direction. If you come to meetings prepared, you'll stand a better chance at shaping your own destiny. It also opens up the dialog for you to learn the next levels of responsibility in the company. It is also helpful to retain notes from these meetings as they will easily compile into a 6-month or annual review document. If you have specific feedback you'd like during a periodic review, it is helpful to send questions ahead of your meeting so your manager's response can be thoughtful.

Network checks are periodic meetings (monthly/biannually/annually) held with mentors, mentees, peers, and other support folks inside and outside of the company. Networks take some energy to create and maintain. Networks are also one of the first to fall by the wayside when crunched with daily deadlines. Resist this. Networks are the singular power for you to maintain a perspective outside of work and are critical to life balance. I keep some formality to some network meetings. I like to have specific questions sent ahead of a scheduled meeting time. Not only do I do this with mentors, I also do this with mentees. Mentees typically enter the network relationship with some hesitancy and an overabundance of humility. I tell mentees upfront that this is a dialog and I expect them to help me as much as I help them. Mentees give me valuable advice on my coaching style, link me with opportunities and other networks, and improve my technical skills. For example, I was being asked by management one summer to consider an external technical expert in academia that I hadn't heard of. I was at the beginning of reading up on this person's work when I mentioned it to my intern and asked if there was some way to electronically search this person's interpersonal skills as well as technical expertise. She deftly showed me a professor rating site and explained the difference between more reliable reviews vs. a lone complaint. This saved me loads of time and didn't I look technically savvy to my superiors! To the intern, this didn't seem like much of an action, but the outcome was strong.

Anonymous reviews such as a 360° review is a computer-generated questionnaire that generates numerical feedback from a broad group of peers, direct reports, managers, and others within the company. Some companies require these reviews periodically for assessment purposes or for promotion. Some companies assist employees to set these up upon employee request. In addition to a numeric evaluation, there is a direct comment function. Both values and specific comments can assist you in career development. Be cautious of reading too much into specifics, however. Look for trends and plan accordingly. Also, do your colleagues a favor and do not use these anonymous feedback systems to simply complain. It is truly meant to be a constructive tool.

Accountability is critical in a corporate setting. Establish weekly or biweekly meetings with your manager and keep notes. I review these notes every six months and write up a summary. Typically, I request a midyear update with my manager to check if our goals have changed and get a semiformal opinion on progress. This gives enough time to correct course if expectations aren't being met and it also diffuses the potential for unexpected outcomes at a year-end review. The natural force is to push these midyear reviews aside as the daily pressures mount. Resist this. It is in your best interest to have this meeting. This is especially critical when you are aware that there is a suboptimal fit in regards to skills, expectations, or work styles. These meetings will help you identify potential communication or expectation issues early and also help you identify new opportunities. If you've leveraged your informal and formal tools properly, your annual performance appraisal should not contain any surprising information and if it does, since you've kept your records throughout the year, both your manager and people who report directly to you can review prior documentation to see where the communication disconnect occurred. This provides you some insulation from mercurial management moments and again allows you to control your own destiny. At each of these points, consider the feedback and decide on a specific action plan.

Other formal feedback mechanisms that are popular in business are style/type assessment tools such as the Myers–Briggs Type Indicator®. These are meant to assist people to understand differences and develop communication tools suitable for their personal style.

Change

The only factor that's constant is change and if you find yourself in a suboptimal situation, that is the harbinger for change. Find the courage to do so. Change can be programmed, such as a rotation schedule that some companies may have (spend a few years in one division, move to another, etc. until a "fit" is found). Some change is less predictable or expected such as an economic downturn or merger. For example, I had interviewed and been interviewed by companies (it really should be a dialog). I vexed over what was the best fit and after selecting the company I thought was the best fit, I sent my acceptance or regrets as appropriate. On the first day of my new job, freshly graduated from college, the company announced a merger. Really, it was within an hour of my arrival! Well, that changed the scope of the day. I had a sense of panic that all my work to understand the landscape had suddenly been rendered useless and I started to question "Did I just make a huge mistake?" Everything I'd read about mergers wasn't good. But, because I had researched both companies merging and I liked them both, I could see that the cultures could merge positively and the companies were actually not that culturally different to an outsider. These were helpful thoughts. I guess what I learned in the end was that even with the best researching, surprises can show up. It was helpful to be flexible enough to see how things would go and it was helpful to have had researched the companies involved as well as have some basic business knowledge.

Career Maintenance

You may have initially developed your professional package as you were interviewing for employment. By professional package, I mean your communication of technical, interpersonal, and leadership skills. Maintain that package by annually reviewing and updating your resume. This will allow you to take a longer view of your professional development and allow you to more clearly set future goals. You may also notice trends as you compare historic resumes.

Every six months I evaluate "goodness of fit" based on my professional goals, the goals of the company, and critical decision criteria. Critical decision criteria are those few essential needs of employment and not a want or wish list. It is a shortlist of deal makers and breakers. My critical criteria for a company are: commitment to diversity, hunger to be number one in the market, tangible employee investment, and clear communication. Embedded in these criteria are the metaconcepts of work–life balance, honesty, acceptance, and practicality. A healthy company invests in its employees without "breaking the bank," knowing that hiring new employees is costly and turnover ultimately dilutes a company's ability for success.

Individual Development Plans (IDP)

Independent of job assignments, IDPs are useful tools. Originally, the concept of the IDP was separate from company development needs and focused solely on the individual employee. This gave it real strength as a tool to retain employees and as a guide for open communication. It has become a growing trend to knit the IDP with corporate goals. I feel this misses the point as a means for open communication, although it still functions as a retention tool. A successful IDP has three elements: goal design, delivery, and company support. Adequate goal design reinvests in both you and the company. It is not tied to project goals and has tangible and measurable milestones. It takes time and revision to develop an effective, measurable goal. Take this time as an individual, with mentors and with your manager to craft your goals. Delivery is essential to maintain the credibility of an IDP. Hold yourself accountable to your stated goals and seek outside review to measure progress. If you do meet with your manager on this topic, keep this meeting separate from project update meetings or performance review meetings. None of these plans are sustainable if the company itself doesn't support this process with people, time, and money. Support from managers and mentors can take the form of making time, actively listening, and becoming truly invested in your success. And it's not all about you, either. Will you be willing to support a mentee in the future?

Money is also an influencing factor for a robust IDP program. Money comes in the form of supplying employee time to focus on this rather than "getting product out the door." It also can be tangibly linked to management's performance reviews, which are commonly linked to bonuses and raises. The idea must be internally marketed as valuable to the individuals within the corporation as well as a valuable asset to the company.

Mentoring

Formal mentoring programs are helpful to become involved in as both a mentor and a mentee. It is a formalized way to network with persons outside of your immediate sphere and it is a good source for honest external feedback and a safe place to test new ideas. Informal mentorships also commonly arise and are very valuable for external feedback as well as improving your overall knowledge of a company. A healthy mentoring relationship is mutually rewarding. Both the mentor and mentee expand their network and learn new things. These relationships can also ebb and flow as needed over the years or may be of short duration. Formally supported mentorship programs within a company and networked out to the community tend to also philosophically support informal mentor relationships. Try to get involved in this positive cycle.

Corporate Culture

It is critical to become orientated to the corporate culture. Mentorships, as I had mentioned previously, are a great way to become networked and "tuned in" to corporate knowledge. A healthy relationship with your manager is always beneficial and clear communication of expectations is a good first step. Understand your manager's goals and work style preference and communicate the same to your manager. It is also helpful to ask for a networking list from your manager. This helps in two ways: it networks you and it gives you better insight to your manager. Expect to go "back to school" to learn about idiosyncratic terms and preferences. Try to identify common ground and find agreement to stated goals and establish "stretch goals." Outline how you foresee attaining those goals.

Networking is a critical function in business. Good manners and appropriate dress are important facilitators of credibility. Casual dress and conversation style may undermine your efforts as a contributor if they are inappropriately applied. Regardless of what your job is, you will likely encounter persons from other countries. If possible, learn about the country and culture that the person originates from prior to meeting. For example, international meetings, especially in Europe, have a more formal sensibility than in the United States.

Specialized Training

Specialized training is critical for both an individual's and company's success. Commitment to safety and adequate technical training demonstrates clear valuation of employees. A healthy company will provide access to improve scientific and technical knowledge of its employees. This can take the form of internal universities or schools or training sessions specific to core business needs. Cereal school, Yogurt U, HR (Human Resources) U, and World Class Marketing Treks are examples of these opportunities within my company. Within each of these schools are a series of classes specific to processing platform or function.

Job Rotation

Personal career development can take the form of in-company rotational development. Some roles are more complex than others within a company and personal preference may be to remain within a few roles or branch to more. Mindful career development takes both corporate need and individual need in consideration to maximize learning opportunities. For instance, one division within a corporation may be very complicated and involve complex processing and ingredient considerations. In this case, it may take longer to gain a working understanding and would necessitate a longer residence within that group. Another division may be new, or in the process of rebuilding. In this case, a more entrepreneurial spirit may be suited. Understanding your key qualifications and learning wishes will help you decide between assignments. There are also transfers between function. For example, quality and regulatory may require more travel than fits in your lifestyle. In this case, perhaps a role in development with longer timelines or less travel needs will allow a more comfortable fit.

Business Organization

Small businesses typically organize their employment model via role: developer, quality assurance, and marketing. Larger businesses typically organize according to role and brand or category. For example, there are separate divisions for cereal, baking, yogurt, etc. and strategic research groups focused on key technical platforms and/or supporting key business divisions. It is important to know how a business is organized so that you can identify who are decision makers or extended team members. This will help you organize your own communication by knowing when and whom to communicate with. This will also help you define what your role or roles are. Smaller companies will likely require you to have a broader technical skill set than a larger one. Larger companies will likely require you to communicate to different groups and manage through others to accomplish tasks.

Travel

Travel is an expected part of doing business. You may be away for a few days or months at a time. This inevitability is a test of work–life balance. Only you will know how much travel is appropriate at the state of life that you are at. It is often helpful to have some routine by planning exercise time. It is also helpful to outline ahead of time what you're willing to pass on in your personal life to fit in frequent or extensive travel. It is also customary to go out to dinner with colleagues or clients. Each company may have a policy as to who pays the bill. Typically, the most senior person or the coordinator/host would arrange details and pay the bill.

Extended travel (weeks–months) is typically due to plant trials, start ups, joint ventures, or potential mergers. You may be part of or leading a team that consists of people from your home office and from the offsite location. It is helpful to build a team before the actual work begins and to maintain this team building throughout your time there. Pre-event meetings are a great way to iron out details and build familiarity. The initial meeting can be a phone conversation, followed by a pre-event meeting with all members physically present at the off-site location. At this time, check last minute details and iron out some contingencies. Before leaving the site, it is best to recap the learnings and send out a summary for all to review and agree upon.

Occasionally, technical meetings or client/supplier meetings will require travel. Because in these situations you are representing your company to an external group, be very cautious of what is disclosed. Protecting critical company information, while still being helpful and collaborative, is a gentle balance.

Proprietary Information

There are three tools to a company to protect their discovery: trade secret, patent, publish (public disclosure). Trade secrets are just that, secrets that your company has documented and protected. A product formula would be an example of a trade secret, that it would be critical not to disclose any information regarding. Competitors can go to great lengths to dig for information. Once, I was part of a project team in a large corporation that was on a really high-profile project that just launched. It caused a buzz in the industry and no one could figure out how we technically achieved success. One day, the phone rang during our Monday lab meeting. The voice on the phone said "Hi! Say, I realize you may not be free to talk right now, but I'm from X company and would love to offer you or anyone on your team a better employment. Just say yes if you're interested and I'll call you at your office phone later." I was astounded. Now, that was brazen and I checked it out to confirm it wasn't some internal prank. Sometimes information mining occurs. Call your security if you've the slightest concern. Unacceptable data sharing can easily and accidentally occur due to colleagues leaving for other competing companies or relatives who work for competing companies.

Patents protect companies by ensuring that they can use a particular discovery and prevent others. Patents are not forever nor are they iron-clad. They also teach others something about your company and may lead to other discovery. It is a way to communicate with others while protecting one's own discovery. One discovery an inventor will have is that just because it is a good idea or a unique idea; it is not necessarily worth patenting. Patents are costly and typically applied for in accordance to business need.

Publishing is another way to disclose discovery. This is not as commonly used as a patent, mostly due to cost, delayed disclosure, and lack of ownership of the discovery. This last approach may be best suited to broad topics in refereed journals, philanthropic efforts, or university-linked discovery.

Product Lifecycle

A product development lifecycle starts with a business strategy. These start at the top as a comment such as "grow share by X% in this category." This is followed by ideas/concepts that are internally developed, often by cross-functional teams. After these concepts are tested with consumers or otherwise narrowed down by other means (e.g., manufacturing cost, availability of key technology), prototypes are developed. These prototypes will also likely be consumer tested to understand what the key factors are for this prototype's success. If the consumer demand meets the technical feasibility and affordability, then a prototype begins scale up/commercialization. In this last phase, the lead developer is responsible for the fidelity of reproducibility from prototype to mass-production. A product's timeline or time allowed to develop a new product or improve an existing one, depends on the market trends, the company's expectation for the product, and desire to market the product. Photo shoots and packaging flats, for example, take time away from product development time. One time, I was developing a new product and my deadline for finished product samples suddenly was moved up by two weeks. The photographer (in Spain) needed "personal time" with the product for an adequate still shot. The resulting photos looked great, but it did make me wonder. New processing platforms can also limit a team's ability to "tweak" a final formula.

Scale up is a critical consideration when developing a product. Typically, development work starts at the bench top and increases in scale to a pilot plant and finally to a manufacturing plant. This evolution coincides with marketing and committed resources. As prototypes are developed, shared internally and tested with consumers, expectations begin to form in other functions on market fit, profitability, and manufacturability. The product will start to "come to life" and this naturally narrows a team's ability to remold it into something else like a different flavor or texture. Timing may also be constrained by external need. Change in federal guidelines or consumer demand for change can severely limit a developer's bench top discovery time. Depending on division within a company, a team may be accustomed to development times of 24 h to years. However, more typical development times for a new product are 6–9 months.

Another item that will impact a product's timeline is documentation. Learning how to efficiently maneuver through this will take a few months; so as a new team member, it is worth your while to concentrate on these details at first. Introducing a new ingredient, for example, will require you to create a new code, have this approved, and put into the database system to communicate to the plants. Each step will require input from many functions and this is the real time robber. So, consider your new ingredient in the light of this to balance the benefit (cost, technical advantage, sourcing).

Safety

Typically, a company will have safety training. This may be valued differently at different companies, but it is critically important to you. Safety training typically includes among other things, how to properly use equipment, waste disposal, and identifies rally points in case of emergency. Be aware of your surroundings. Are there any hazards in your work area? (Look for tripping hazards, overloaded circuits, etc., and contact your safety person rather than letting it go.) Know where to go for a fire evacuation, severe weather, or other occurrences (e.g., earthquake). When traveling, always share your itinerary with your manager or administrative assistant. In the days immediately following the 9/11 attacks, my company did a roll call to determine if everyone was safe. This was made easier by other's having ready access to travelers' itineraries.

No deadline is worth getting hurt over. Lack of sleep leads to poor judgment and that can lead to disastrous outcomes. It is one thing to study all night and try to take an exam. Only your grades will suffer and you can probably make up the sleep the next day. When working with manufacturing equipment, the consequences can be more severe.

Remember that as a representation of someone outside of the manufacturing facility, always use good safety practices, even if you don't see the plant itself following that practice. For example, there typically is a pedestrian door next to the forklift door (large roll up door for motorized traffic). Take the pedestrian door, even if it is plant habit to just walk through the large door. Not only does this maintain that you value safety, you don't know the details of the plant (When/where do forklifts come by? How fast do the drivers go?). It is also a good idea to travel with your own personal protection equipment (PPE). I know plants usually keep these things on hand, but your stuff will be more comfortable. I also include a safety section in my pre-plant meetings to orient myself and the team I travel with. It sets a valuable tone that I don't wish anyone to be injured during a production run.

Contract Facilities

You may find yourself traveling to a contract facility, another company that your company contracts to make product for you, rather than a company-owned plant. Dealing with a contract facility requires slightly greater skill than dealing with people within your own company. You will have less control over pretrial meetings and you will need to screen your critical information prior to entry. I try not to bring my laptop to contract facilities, but rather print out critical sheets ahead of time and will use paper and calculator rather than an electronic spreadsheet. A heightened sense of corporate security is necessary as the contract facility's level of corporate security may not equal the security expectations of your company and you also stand a greater likelihood of running into competitor's products and employees. Your access to parts of a contract facility will likely be blocked and it is good practice not to wander alone within a site. You may also have the unexpected pleasure of running into an old school friend who works for a competitor! On the one hand, you never want to be legally responsible for knowing something proprietary about another's company. On the other hand, it is rather nice to see a friendly face.

Realize that a contractor may have a very different value system than your company and try to be diplomatic without compromising core values.

For example, I was joining my team late one time for a plant trial at a contract facility. I wore my PPE on the flight over (imagine the odd looks) and was peeling off my coat as the administrator showed me a long hallway of executive offices and told me which office was ours for the day. As I went down the hallway, I overheard a conversation. "Yea, there's one more coming, some guy named Dr. Chris something. Long last name. I thought tonight we can take the guys out to that strip club and talk business. You know, let the conversation and juices flow. Good for business. Yea, reserve a table." He hung up. The door I was directed to was the one just past his. I poked my head in his office and pointed to the name on my shirt. "Hi. I'm Dr. Christine Nowakowski from...company. Um, sorry, but I couldn't help overhearing your conversation. Thank you for the kind social invitation, but I'll have to decline." He didn't say much, but managed an embarrassed, "Hello" and shook my hand. I went next door to plunk my stuff down and meet up with the rest of the team. I heard the guy pick up the phone again. I had my pretrial meeting as usual and the run went fine. Oh, and dinner was very nice too.

Legal

Knowing the law and your company policy are important tools for communication during interviews. Know what is appropriate conversation and questioning per your policy. If your company doesn't have the resources for this, review a basic business law textbook or take a class. Some things to know are what a protected class is, what appropriate things to say/do are, and what is inappropriate. Know also next steps if inappropriate behavior has been witnessed. Get human resource (HR) department involved early even if it is to ask questions that come to mind or things that you thought in a "gray area" of appropriateness.

Technical Communication

Because of timelines and diverse functions, communication is very important. Knowing how to relate concise, accurate technical communication at the appropriate time is critical for a successful career. Two sentence summaries are common as are short presentations including data sheets and other supporting materials. It is important to have a short summary of your activities that can inform a variety of functions quickly. For product development, typically external functions wish to know what product, when, and what specifically is their role in the development process. Internal functions will want to know what and when, but also how you're developing the product. This type of sharing can shorten development time and lessen effort. This is especially true if similar products are being developed or different products from the same manufacturing platform. It is helpful, when framing a response to an information request, to envision what the recipient would like or need to know. The vice president may be more interested in positioning strategy in the marketplace than the actual details on how the product was made. Technical personnel may be more interested in the details of formula and processing conditions than positioning strategy in the marketplace.

Communication is important to avoid internal competition or task redundancy. Having two people doing nearly the same thing is wasteful of resources and is deflating to those working so hard on something. Communication is critical at all levels of an organization to avoid this pitfall. The likelihood of redundancy is higher in large corporations, especially when a project is still in its early phases. Communicating out and often to your network will help others know what you are working on, and therefore, will refer to you when asked about a particular project. If you discover someone is working in a very similar area, communicate early with your peer as well as with your supervisor to see how a collaboration or realignment of goals could be accomplished.

Managing Others

As you get further along in your career, you will gain responsibility from project manager to supervisor with people reporting directly to you. Take the time early in your career to learn as much as you can about the next level. Seek out mentors at the management level and start reading those business books! It will help you streamline communication with your manager by knowing the lingo and knowing the expectations of their job. It will also help you demonstrate success before the actual responsibility comes your way.

Overcoming Resistance

Resistance can come from internal and external sources. Pet projects, perceived competition for promotion, and in general, lack of understanding can all get in the way of progress. Frequent and varied communication is key to overcoming resistance. Try to understand others' points of view and carefully consider how you can incorporate their concerns in your vision.

I've heard in manufacturing locations that employees from corporate headquarters are sometimes called "seagulls" (they fly in and poop on everything). People were joking at the time, but realize that when "corporate" goes to a plant, you are representing not only yourself. This goes the other way also; you may find yourself as an employee at a plant location, interfacing with "corporate" past history with others may strongly influence your dialog within your current project. It is very useful to know what the corporate relationship is with a manufacturing site. Try to travel with a team, especially if it is your first trip. Also, try to find out who has been to a location before to get greater detail, not only on things like process platforms and layout, but also who would be good "go-to" people. Typically at a manufacturing site, there is a system engineer, quality engineer, or plant manager who can help you either directly or point you to a line manager or team leader that could help you. Gather these names and keep a file back in your office. They will become part of your extended network that you can leverage in the future. It is helpful if the plant buys into your vision and it is worthwhile to plan an in person preplant meeting prior to any manufacturing run.

Measuring Success

Occasionally, you do everything right and still, a new product or project fails. I have worked on many products that have been discontinued. Rather than seeing things in a go/no-go light, think about a value captor model. In this model, strategic boundaries are set, key areas of opportunity are identified and progress is monitored. If a project is "successful" in the go/no-go model, it will launch to a final product. If it doesn't launch, it may be "killed" or "dead." In the value captor model, there are more outcome options: spin product learning into the original project to modify goals, spin off to another project, or salvage learning/product to retain for another day. For example, a company invested heavily in a manufacturing platform to make intermediate moisture dog food in the 1970s. A product was developed, launched, and was a remarkable failure. However, the division responsible was stuck with millions of dollars of equipment that was not being used. By salvaging the newly acquired platform knowledge, developers quickly got to work to find out what could be manufactured at a profit. The result was a highly successful (and profitable) food product that exists to this day.

Summary

I think success is building toward a long-term legacy (the tangible outcome of personal skills) and a lasting positive network (who you know to get things done). Over a span of a career, you'll work with lots of different people, likely work in different companies, and have lots of failures, successes, and above all, have accumulated great stories. When things get challenging, always try to keep in mind what your core values are and never compromise those. Yet, try your best to be as flexible as you can be when interacting with others. Finally, be open to surprises and have faith in your and your team's abilities to get things done.

Part V
Careers with a Degree in Food Science

Chapter 18
Quality Assurance/Quality Control Jobs

Melody Fanslau and Janelle Young

Quality is never an accident; it is always the result of intelligent effort

John Ruskin

The production of a quality and safe food product is essential to the success of any food manufacturing facility. Because of this great importance, a career in quality can be extremely rewarding. Without happy customers willing to buy a product, a company would not be able to survive. Quality issues such as foreign objects, spoiled or mislabeled product, failure to meet net weight requirements, or a recall can all turn customers away from buying a product. The food industry is a customer-driven market in which some consumers are brand loyal based on a history of high quality or in which a single bad experience with a product will turn them away for a lifetime. With this said, the main role of a quality department is to help ensure that quality issues such as these are eliminated or kept to a minimum to maintain or increase the number of customers purchasing their product.

Being a part of a quality department requires a set of distinct characteristics in your quality tool belt that will enable you to successfully solve problems. Problems that arise can range anywhere from employee relationship problems to manufacturing issues. Sometimes you will have to solve a problem quickly, for example, to keep a production line running while other situations will allow you ample time to fully research and develop your solution. In both cases, it is an extremely rewarding experience to successfully solve a problem and see first hand how your solution helps a company produce a safe and quality product for its consumers. You may already

M. Fanslau
Fair Oaks Farms, LLC, Pleasant Prairie, WI, USA

J. Young
Assistant Quality Manager, Lactalis, Belmont, WI, USA

R.W. Hartel, C.P. Klawitter (eds.), *Careers in Food Science*,
DOI: 10.1007/978-0-387-77391-9_18, © Springer Science+Business Media, LLC 2008

possess some of the tools you will need to be successful in quality and the others you will be able to develop while in the field. Overall, a career in quality is an outstanding way to find your niche within the food industry while developing a skill set that will aid you in all future endeavors—whether they are in the food industry or within another business sector.

Skills of a Quality Professional

To be successful in the quality field, it is necessary that you be, first and foremost, a good listener. It may sound simple, but good listening skills are key to analyzing a situation and determining a proper course of action. Being able to listen and get a good handle on all sides of the situation before reacting will help you to make a sound decision. Being able to communicate well goes hand-in-hand with listening. As part of a quality team, you will be called upon to make decisions and communicate those decisions to others. Oftentimes, someone may not agree with your decision and you will have to explain your stance in a clear and concise way in order to facilitate cooperation. The “someone” that quality professionals typically find themselves in disagreement with is production personnel with the nature of disagreement following the age-old dilemma of “quantity versus quality.” The fact of the matter is that any business cannot survive without sustaining the delicate balance of producing a large volume of product (quantity) while producing a good wholesome product (quality).

It is for this reason that effective communication skills are an important tool to always have in your belt. You have to understand that not everyone is going to see your point of view in every situation, so being able to back up your actions with logic while understanding where the gray area lies is essential. For instance, you will probably not shut down a production line for a couple of mildly skewed labels—the product may not be the “ideal” end product, but in the end, it is not likely to affect customer perception of the brand and prevent future sale of the product. Shutting down the line will however hinder production, and the success of a business depends on the efficient production of product. Nevertheless, there are instances when shutting down the line, no matter what the cost, is vital. In any potential food safety issue, it is important to act immediately whether it is stopping a production line, putting a product on hold, or shutting down a production room. You can always resume operations or take a product off-hold once the potential food safety issue has been cleared. In the end, it is essential to understand the balance between “quantity versus quality” in order make both the company and consumers happy.

Another skill that is valuable to have in your tool belt is the ability to multitask. A manufacturing facility is a fast paced and exciting environment.

Because of the speed at which production occurs and the need to get quality product out of the door, the quality department is often called upon to do numerous tasks at once. The ability to multitask in this fast-paced environment is a valuable skill. You must be able to determine which task has the highest priority, tackle that one, and then move on to the issue with the next highest priority. For example, you may be in the process of conducting one of your required production line checks when it is brought to your attention that there is possible metal contamination on another production line. Due to the importance of taking control of any food safety issue such as metal contamination, you would cease conducting your line check and tend to the food safety issue. Once the food safety issue was taken care of or under control, you would go back and complete your production line check.

The ability to work with a cross-functional team that will most likely include people from all levels of education and industry experience is yet another useful skill to have or to develop. In a food manufacturing environment, you will likely get a chance to work with production, maintenance, sanitation, product development, and management. In most cases, you will not be able to correct a problem or improve a process without the help of one or all of these individuals or departments. Being able to work together with a team is a key to solving most problems that can arise in a manufacturing environment. Many food manufacturing facilities are going toward a Total Quality Management system (TQM) where it is typical to have cross-functional teams solve problems. There are multiple benefits to a TQM system including a greater sense of ownership in the process, more efficient systems that work for multiple departments and creative ideas that may come from unexpected sources. For instance, at one food manufacturing plant during a preventative maintenance troubleshooting session in which a TQM team was assigned, a marketing manager who had never wielded a grease gun came up with a solution to the overgreasing problem that was occurring throughout the plant. This was something that saved time (less equipment shutdowns due to malfunction), money (cut down on unnecessary costs), and product (potential contamination of excess grease into product).

While each of the aforementioned traits may be characteristic of someone choosing the quality profession, everyone brings their own set of ideals and experiences to the table. There are many different positions that fall under the quality umbrella and as such, there is a little something for everyone to explore. The structural hierarchy of the quality profession within the food industry will vary based upon the size of the organization with which you are employed. Following is a sampling of a few generic positions in which college graduates entering the quality field may find themselves employed.

Quality Control/Quality Assurance Positions

Laboratory Technicians

Laboratory technicians generally are the go-to people within the quality department that provide support through the generation of data. Technicians may conduct various microbiological or chemical tests that may be used for raw material inspection, product release, Certificate of Analysis (COA) generation, or problem solving in the event of a deviation in the process. A COA is a document provided by a supplier to ensure that the testing required by the customer is within the set specifications. This testing is usually done on a "lot" basis to represent anything from a single batch of product to an entire production day's worth of sample. This may include information such as percent carbohydrate or total bacterial load. Some companies may use this document in support of their food safety plan or in process control to ensure that the product is safe and consistent each time it is produced. Depending on the organization, a technician may also be involved in shelf life or sensory testing of a product. In addition to product testing, the technician may also be responsible for environmental monitoring to provide data supporting the plant's food safety program. The typical work atmosphere consists of a laboratory setting with periodic work out on the production floor.

If this type of position is something that you may find of interest, nothing is a better preparation than working as a student laboratory assistant either on campus or within the private sector. Many professors look for extra help within their laboratories and are willing to take the time to train and aid in the learning process. While ideally you would want to gain this employment in the field of food science, it is not necessary in order to get an understanding of laboratory practices. For instance, you can work in a plant microbiology lab and gain experience in sample collection, plating, sterile technique, and other good laboratory practices (GLPs) that will be both impressive and desirable to a potential employer. In addition, it is a good idea to gear your electives toward chemical or microbiological courses including lab work to gain greater experience and understanding. Also, since this type of position usually involves a lot of data collection and analysis, it would be beneficial to take advanced computer courses in database programs to gain proficiency in data organization and manipulation.

Line Auditor

Line auditors are responsible for monitoring the daily activities on the production lines. This monitoring function usually includes parameters such as net weights, package integrity, product code dates, and overall product appearance. In addition to package or product monitoring, the auditor is often involved in the monitoring of GMPs (Good Manufacturing Practices) including employee hygiene and sanitary line conditions as well as conducting facility audits such as pest control, release of equipment or rooms, and general plant inspections, which identify potential hazards throughout the plant. All of these monitoring functions are conducted to ensure the compliance with the quality standards set by the company or by regulatory agencies such as the United States Department of Agriculture (USDA) or Food and Drug Administration (FDA). A position as a line auditor will allow you to develop your communication and troubleshooting skills. You will likely be required to resolve manufacturing issues such as noncompliant product or sanitation problems through the interaction with production supervisors and line workers as well as with quality and plant management. A quality auditor position is a great place to start out in the food industry to get hands-on experience out on the production floor and gain a better understanding of the process as a whole.

Many internships within the food industry are auditor type positions, so it would be advisable to seek an internship if this position is something that interests you. As mentioned above, the line auditor has a lot of interaction with sanitation and production. To gain knowledge in these areas, it may be beneficial to take some microbiological courses aimed at bacterial growth and inhibition. These will allow you to present your stance in a clear and educated manner in the event that a line needs to be shut down due to a sanitation issue or potential contamination issue. In addition, there are organizations that may offer courses in foreign material handling and assessment as well as public health. Both of these will help during line audits. If you have a mechanical aptitude, this may be a career to pursue as a stepping stone into a process improvement or quality engineer type position. In this case, mechanical engineering or similar type courses may help you achieve your career goals.

Supervisors/Managers

In general, a quality supervisor or manager position requires someone with good understanding of the process as a whole along with good multitasking and delegation skills. The supervisor will be responsible for reviewing the quality process, facilitating training activities, and communicating the quality philosophy of the plant. Using key performance indicators (KPI), the supervisor or manager will identify positive and negative quality trends to determine which sectors need the most attention to continually improve the process. If pursuing a career as a quality supervisor or manager, it is important to keep in mind that each day is constantly changing. While there are some routine activities that need to occur, a person in this position needs to be able to prioritize their activities based upon the daily needs of the production environment and the support staff.

Typically, this position is one that you need some experience working as a laboratory technician or a line auditor before acquiring. Internships or prior experiences working in a similar capacity may allow you to enter this profession without first starting off as an auditor. That being said, it would be beneficial to take similar type of course work as you would for a technician or auditor. In addition, due to the nature of a management position, it is important that you have effective communication and presentation skills. Quality supervisors and managers typically will have to conduct trainings at either the department or plant level, so public speaking skills are essential. To better hone these skills, take some technical presentation and writing courses so that you can present your quality findings (such as KPI) in a concise way that makes sense to those you are presenting to. If you are confident in your speaking ability, it will show and you will be more successful at getting your point across.

HACCP Coordinator/Food Safety Specialist

A Hazard Analysis Critical Control Point (HACCP) system is a food safety program that focuses on the identification, evaluation, control, and prevention of hazards at all stages of the food production process. A HACCP plan is required of all USDA inspected plants; however, many FDA plants choose to develop HACCP plans as well to identify potential areas within their process that may pose a threat to the overall food safety of their product. Not all companies will have a HACCP coordinator on staff. Oftentimes a quality supervisor/manager will take on the role of managing the HACCP program and all food safety issues.

HACCP coordinators are responsible for maintaining food safety standards of a company. They are the liaison between the regulatory agencies and the company. In some cases, a HACCP coordinator will have to develop a HACCP program from scratch. This requires development of the foundation of any HACCP plan, prerequisite programs. Prerequisite programs can include set standards for transportation, storage of raw materials, sanitation, GMPs, and pest control. Once prerequisite programs are established, a HACCP coordinator will develop a HACCP plan using plant history and scientific evidence to support their decisions. After a HACCP plan is developed and implemented, maintaining a sound HACCP program becomes the main focus of the HACCP coordinator. Maintaining a HACCP program includes keeping up-to-date with new government regulations, continuous training with current and new employees, and constant monitoring of the program to ensure justifications used for initial decision making still stand. Overall, it is the responsibility of the HACCP coordinator to maintain a working HACCP program that strengthens the food manufacturing facility by helping to ensure the production of a safe product for consumers.

HACCP coordinators or similar type of positions should have a strong background in the three types of hazards addressed by a HACCP: bacterial, physical, and chemical. Advanced microbiology courses will give any future HACCP coordinator a good skill set at assessing the bacterial hazards. In addition, physical hazards pose a threat and coursework or experience in foreign materials and public health will help for this position as well. Chemical hazards may include sanitizer solutions or chemicals or something such as allergens that pose a chemical threat to the end consumer. Knowing how chemical reactions may affect product quality and consumer safety are two important skills that any HACCP coordinator needs in their tool belt. Regulatory know-how is also essential to understand what rules and regulations have to be in place to be in compliance with your plan. Many universities will have food law courses that can be taken to gain this experience. Like a quality supervisor or manager, it is also essential to be a good communicator both in written and oral forms.

As mentioned earlier, the hierarchy of an organization varies from one company to another and the level of education required to perform different tasks is dependant upon this hierarchy. For example, larger food companies typically will require a college degree for line auditor or technician positions, while smaller ones may have these positions filled with a high school education. When looking for a position, it is important to look into what you want to get out of the experience and make your decision based upon your goals and aspirations.

A Day as a Quality Professional

Now, you may be asking yourself "How do all of these positions work together to assure a quality product is being produced?" or "What can I, as a quality professional, expect from a typical day on the job?" Let's take the following scenario: The corporate office is made aware of a problem with Product X, a new product that has been on the market for several months and is now getting customer feedback. Unfortunately, it seems that some of the customers are experiencing severe stomach problems after consumption of the product. Time is of the essence, so it is important to employ cross-functional teams to determine how much product is potentially affected, what about the product that may be causing the adverse reaction, and what is the final course of action to segregate this problem and prevent a potential recurrence. While this scenario is not a common one, it is one that quality would be expected to solve.

Generally, the quality manager and plant manager would be the first to be made aware of the issue and would initiate the investigation by employing cross-functional teams to determine what occurred during the production of Product X and how it can be eliminated. To start, the quality manager would facilitate the investigation by determining what lot was affected or what production date was involved and try to get samples of this product returned. With this information, the food safety specialist, quality auditors, and laboratory technicians can begin to narrow the scope of the problem. The food safety specialist will begin to investigate any potential concern by reviewing microbiological data for the product and environment along with the quality technician. Shipping and receiving and production will need to be called upon to work with quality to determine how much product was potentially affected and which customers received the product in question. Quality auditor's reports will be reviewed so that any potential production issues that may have occurred during the process can be eliminated or verified.

With all of the resources and people involved in the quality team, a resolution can generally be made in a timely fashion, but it requires the cooperation of every person and their unique responsibilities. No single position can be solely in charge of assuring the quality of a product, which is why teamwork and communication skills are vital to the success of any food manufacturing facility. Each department within the organization may be a separate unit, but all activities can be tied back to the quality department. That is what makes a career in quality so exciting. It gives you exposure to every part of the production process and allows you to develop valuable problem-solving skills.

What You Can Do to Get Started as a Quality Professional

To better help you achieve your career and personal goals, it is often helpful to look to different organizations or certifications that will give you an edge by providing you with a greater skill set and working knowledge of the food industry. Most professional organizations have discounted membership rates for student members, so it is important to take advantage of the knowledge that can be gained from these organizations. In addition, many of the organizations offer certifications and courses that can either be taken online or at a predetermined location. Some examples of certifications that food manufacturing facilities may look for when hiring someone to fill an open quality position are HACCP, ISO 9000:2001, Lean Manufacturing, and Six Sigma Black or Green Belts. There are also more industry-specific certifications that you may want to look into based upon your individual interests or career plans. Following is a summary of some of the general certifications that will aid anyone interested in a career in quality and will help in other sectors of the food or manufacturing industry.

Certifications

HACCP (Hazard Analysis Critical Control Point)

HACCP certification is offered by many different organizations and can range from a very basic course that gets one familiarized with the plans and what the different requirements are to more intense courses that prepare a person to be a facility HACCP coordinator. If you think you may be interested in pursuing a career more along the food safety side of quality, a HACCP certification is probably right for you.

ISO 9000:2001 (International Standards Organization)

ISO 9000:2001 is more of a manufacturing tool than a quality specific tool; however, many food industries are ISO certified to provide their customers with a greater level of confidence that they will be receiving a product produced within specification every time. ISO is a system of checks and balances that makes sure that a plant is doing what they say they are doing, through careful document management and maintenance. This becomes especially important in the quality technician or quality auditor roles where monitoring and measuring of product and processes is an essential duty. Irregardless of whether a food-manufacturing plant is ISO 9000:2001 certified, the skills learned through the ISO certification process are excellent ones to develop from an auditing and investigatory standpoint. These courses can be offered through local technical colleges, through Department of Labor subsidized programs or through the ISO Web site.

Six Sigma/Lean Manufacturing

You may have heard a lot of buzz about Six Sigma or Lean Manufacturing. Both of these manufacturing philosophies have their roots in eliminating unnecessary loss throughout the process while continuing to provide a quality product to the consumer. Plants that practice one of these manufacturing practices generally value a TQM system as well and have their quality departments heavily involved in cross-functional teams. Obtaining certification as a Six Sigma Green or Black Belt will help you to "knock out" the competition when looking for a position in quality.

Organizations

ASQ American Society for Quality

The American Society for Quality is an industry-wide organization devoted completely to quality. They are a great resource for gaining a greater understanding of quality principles, problem-solving tools as well as up and coming quality initiatives. In addition, ASQ has a comprehensive training and certification program including quality auditor/ technician/ engineer, HACCP auditor, and many others. There are multiple divisions that you can get involved in based upon your interests and potential career path.

Institute of Food Technologists (IFT)

IFT is a great way to network and get more information about the trends and happenings within the food industry. In addition, there is a great student organization that provides an individual the opportunity to get involved in leadership activities and various poster presentations, product development competitions, and other student activities that will aid any student in gaining experiences valuable to an individual who is thinking a career in quality may be right for them.

In addition to the aforementioned general organizations that a potential candidate for the quality profession may belong to, there are always industry-specific organizations and organizations within the university and college that you may attend. Further, the quality profession is getting a lot more streamlined in terms of technology. It would be beneficial to gain an understanding of graphing programs such as Microsoft Excel or basic statistics, as these are some of the tools used to track the cost of quality within an organization.

Summary

The Quality profession can be an extremely rewarding career for any one interested in being an integral part of the production of a safe and wholesome food product. Even if you are not exactly sure what area of food science you want to focus your career on, a position in quality can be a great place to start. While in school, it is important to take advantage of the resources and opportunities out there. Talk to your professors and advisors about what you are hoping to accomplish after graduation and they can help point you in the direction of internships or jobs that may help you attain your career goals. In addition, student organizations are a great way to network and find out more about what interests you in your quest as a quality professional. Food Science clubs often have guest speakers who talk about careers and general principles of the food profession. Take courses in areas that will help you gain the skills needed to be successful, but most of all, have fun with it. Learning is much more rewarding when it is enjoyable. By immersing yourself in your chosen field, you will feel more connected to it and preparing for the "real world" will seem a less daunting task. By putting forth an intelligent amount of effort, you can ensure that you are prepared for a rewarding career in quality.

Chapter 19
Production Management

Richard Boehme

Introduction

One of the most overlooked jobs by a new food scientist graduate is in production management. People in production management are responsible for the day-to-day operations in a food plant, where raw materials are turned into finished goods for a customer.

The first step in production management is a production supervisor. A production supervisor is the one directly interacting with the plant employees. When employees have a question or an issue comes up, the production supervisor is the first management person to attempt to solve a problem. They will constantly be in the production areas monitoring production rates, product quality, employee interactions, and anything else related to the plant that may come up. They may also specialize in a certain area of the plant, like safety, sanitation, or a certain production line, and look to make improvements to that department.

A production manager sees more of the big picture. They are able to plan ahead and look toward the future. They will work with the production and employee schedule, hire employees, and look at executing longer-term projects. They will also help the supervisors with any complicated issues that may come up.

There are numerous positives of being in production management (PM). You are in charge of making the products your company sells. All other jobs within the company should support you in your job. The work in research and development (R&D) would be useless if the products they design can't be made correctly and efficiently.

R. Boehme
Kerry Ingredients, Beloit, WI, USA

R.W. Hartel, C.P. Klawitter (eds.), *Careers in Food Science*,
DOI: 10.1007/978-0-387-77391-9_19,

A Day in the Life of a Production Supervisor

A production supervisor directly oversees production employees at a food plant to produce finished goods for customers. This could include product running on multiple production lines, many different finished goods for a wide variety of customers, new products being scaled up by R&D, to name a few. The production supervisor's job is to make sure all of this happens as efficiently as possible.

There are two main areas that a supervisor has to deal with, (1) understanding the finished goods being made with the raw materials used, and (2) effectively leading a team of production employees. Production supervisors may come from a variety of backgrounds; from a food scientist to business management, ex-military, accounting, and engineering. A food scientist would have an advantage in that he/she should know how food components interact and will understand better how to troubleshoot any issues that might come up. A food science degree gives a definite advantage to the product supervisor.

Since one of the most important job responsibilities of the production supervisor is managing people, it is important to know something about the people you would be supervising. There are two different sets of employees—union and nonunion. A union is a group of employees who bargain with plant management to improve working conditions. Most plants will have a set of plant rules to help protect their interests, which include having a reliable workforce that will show up on time and perform their work duties to help the business succeed. The union will most likely have a contract with the plant that will protect things in their interest, including having a steady pay, vacations, and other rules specific to that plant. Every 3–5 years, the union and plant management will negotiate changes to the union contract.

Although every day will be slightly different, we can look at what might happen on a typical day for a production supervisor. The start of the day will be a shift change with the previous supervisor. From there, you'll figure out what happened in the last couple of hours, what products you are running right now, what you will be running later in your shift, and any issues that may be coming that day. This is a good time to ask questions. Sometimes during a hectic day, the other supervisor may have forgotten a little detail that could cost you valuable time to figure out. Also, it may be a good idea to take notes as you also may be running around all day and forget something you learned at the beginning of the shift.

After the shift change, it's time to get on the floor and figure out what's happening. This is when you'll catch up with the employees working that day. Although the supervisor you relieved should have a good idea about what's going on, you probably can get a lot of valuable information from your employees too. You will want to make sure the expectations are clear to the employees, in case there is anything odd going on.

After you make your rounds, and everything is going well, this is the time to work on special projects. If you want to advance your career in the food industry, simply being a supervisor and just making good product on your shift isn't enough. There are always opportunities to increase throughput, add efficiencies to other sides of the plant, and make life easier and safer for the employees. To get ahead, you need to show initiative—look around, figure out what needs to be done, and do it (without sacrificing any of your primary responsibilities, of course!).

Some days won't go smooth at all. Let's start off by having two employees call in sick, you're having troubles getting someone else to come in early or stay late. So right at the start, you're shorthanded and shuffling employees around. Your most critical product line for the night is behind its production schedule and you're in jeopardy of missing customer orders, so you will want to spend some extra time with the employees there to make sure they stay on task. Then, you find out from the lab that a different production line is making "out of spec" product that won't be good enough to ship to a customer. You have to figure out why and fix it. The problem could be coming from the raw materials, a broken piece of equipment, employee error, or the line may need to be washed to fix the problem, which will delay the schedule for that line. As a production supervisor, you could have all of these going on at the same time and it can be overwhelming. It is days like those, which will happen, that you just try to keep the ship afloat. Being a production supervisor will take the skills needed to multitask in situations like that. The good thing about production supervision is that, after a while, you will get used to situations like that. Once you're used to the pressures of PM, you can handle nearly any multitasking event life will throw at you.

Tips for Production Supervisor

Here are some tips you might find useful should you choose a job in PM.

- Be prepared to work nights and weekends. Most plants run 24/7 and they may need supervision on the odd hours of the night.
- Take over a department—make life easier for the employees. Taking care of problems that come up is essential to supervision, but if you want to progress in your career, you should look at improving things. Look to lock onto one department and try things like improving their paperwork, cleaning up the area, and making up instructions for operating equipment or other parts of their jobs.

- Be honest—Employees need to trust you. That one time you make a mistake, the employees will remember it for years.
- Listen to your employees—They know a lot. Let's face it, you're fresh out of college and although you might think you know most processes of making food, you don't have the in-depth knowledge for each of those processes. The employees working on a line will be just the opposite. Ask them questions like: Has this happened before? What did you do to fix the problem? How did that work? Questions like that will give you ideas on which directions to go in when trying to fix a problem.
- Find what you do better than anyone else, and show it off. For instance, if you are comfortable working with computers, maybe you could make a spreadsheet that could greatly reduce the amount of time to do a data collection process. This could free up time for another member of the team. Time spent doing redundant activities do not improve the plant, and helping other members of the team will help the plant grow.
- Work as a team with everyone else. You'll need each other. The more you can understand what other people around the plant do, the more efficiently you can make things. Take time to understand how the warehouse works so you can deliver products to them the way they want it. Pay attention to what your data entry people do. They interpret the numbers that everyone will be looking at, and also have a lot of tricks to finding valuable information. Work with scheduling. There are more than likely opportunities in product scheduling sequences and lengthening time between washing.
- Go through your boss on everything. Make sure he/she knows what you are doing, but don't over do it. The better you make your boss look, the better it will be for you in the long run. Also, don't go above them until you've talked to them first. Jumping rank on sensitive issues can be a career-limiting move. Make sure you go through the proper channels, which starts with your boss.

How to Prepare for a Job in Production Management

If you think you might be interested in a job as production supervisor, there are some things you can do to help prepare while still in college.

- Think Big. Things on the lab bench or classroom are only a tiny fraction of what you'll be doing in industry. You can't just replace full production equipment for a couple hundred dollars and put it in the washing machine when it's dirty. Try to think of how you can relate what you are doing in college to full-scale production. For instance, instead of heating up food slurry in a glass beaker on a hot plate, you'll be using a 1000 gallon tank with steam-jacketed heating for multiple hours.

- Get into the pilot plant—learn to run some of the equipment. Tracing the piping can be interesting. See where product comes from and where it goes. If you see a valve somewhere, ask yourself, why is it there? If there's an odd piece of equipment you don't know, ask someone what it is and why it's there. The earlier you can start learning about different kinds of production equipment the better.
- Get into costs as much as possible. Everything costs money. Try your best to pay attention to things people do to reduce the costs. For instance, instead of having an expensive formulation, maybe changing the addition or processing of a formula can give the same effect without the cost. The food industry is always looking to get a cost savings by creative thinking like this.
- Mass balancing is important. It's good to be able to follow the process of adding ingredients and seeing what comes out. In industry, things can get complicated. When you put in 100 pounds of product, you probably are only getting 90 pounds out. Where did those 10 pounds go? If you're adding in steam to heat your batch up, there are extra pounds of water going into your batch. How are you going to account for that? A percent of loss here and there in production can add up to big money.
- Group Projects. You most likely won't get to pick your employees or management team, but you will have to work with them. Working with your fellow college students will be a walk in the park compared to the range of people you'll find in industry. You can go from people who can't read or speak your language to management who are only concerned about their career progress to the person who just floats along and doesn't contribute above their daily efforts to improve the business. Learn how to work with and get along with others, regardless of who they are or where they come from.
- Internships. There aren't many internships out there for production management. What you can do, however, is get a job inside a production facility. A good suggestion is a quality assurance (QA) job inside a plant. Working as a lab tech will give you a chance to learn what's important about a product from a customer's prospective. If your products fail in the lab, they most likely won't be shipped to the customer. Working in QA may also give you a chance to walk around the plant and collect samples. This is where you can begin to see what production is all about. A good tip while being a lab tech is to ask questions. Why is this product out of spec? What is causing it? What are you going to do to fix it? These are exactly the questions a production supervisor will answer during the working day.

Finding and Interviewing for the Job

Finding a job as a production supervisor should be similar to finding other jobs out of college. Networking, Web sites, job fairs, and job postings are all worth looking into. One advantage in production management may be less competition for jobs because they simply aren't the first choice among food science students, perhaps because of the perception that their food science background isn't fully utilized.

When you're going into an interview for PM, try to be as calm and confident as possible. You need to show that you won't be intimidated by others, but also need to show respect for others. These are skills you'll need on the job when working with employees, so you should be comfortable in the interview process. One more tip, if you are interviewing at the manufacturing plant, you shouldn't wear ties or loose clothing, as they can be considered unsafe while moving around machinery. Also, you might not want to wear a skirt since some parts of a plant tour may be on elevated platforms.

One thing you'll need to demonstrate is that you can work with and direct people in a working environment, and that you can learn and improve a process in the food industry. The best way to do this is to show the interviewer previous experiences you've had and what you did in those situations. Most likely, this will come from group work you've done in college. Try to show all aspects of how you participated in all aspects of a team: from being a leader, to following and executing instructions, to motivating others in your group. Also, it's okay that not everything goes perfect in a group activity. It's okay to say how something went wrong, and show how the problem got fixed.

Another thing you should talk about during an interview is a food process you've worked with in college. Be able to map out the process from start to finish. Where did you get the raw materials from, what equipment did you use, what time/temperatures were used in the process, what changes did you make in the middle of the process to improve things, and what were the final results of the project. The more attention to detail here, along with showing problems with solutions, would be very helpful.

A Personal Story

When I was looking for a job, PM wasn't my first choice. I heard from a professor that this company was looking for a production supervisor. I went up to the plant for an interview, liked what I saw, and took the job.

I started out as a production supervisor, spending a lot of time on the floor by the production lines learning the process and the products that we made. I was somewhat a late bloomer; it took me about 2 years before I felt really comfortable with most things. During this time, I developed a "make it happen" attitude, which was very helpful. This means you clearly communicate goals and objectives to the employees. Follow up with them to make sure they know what they are doing, and then keep monitoring them during the process to make sure they are on the task, and can answer any questions they have. This is critical to insure that things get done.

After that, I wanted to show the company that I had special talents to offer them. I took over a department and tried to make them more efficient. The most productive thing I did was set up a weekly meeting with the employees and scheduling to talk about the production sequence. These were very successful meetings as we streamlined the schedule, eliminating a lot of cleanouts and increasing throughputs by double digits without spending any money on improvements. I was also comfortable using computers, so I put together a lot of data for the department that also helped with scheduling.

From there, I focused on product formulation. I was able to optimize formulae from a cost perspective while still maintaining a quality product. This also tied into inventory control. While working with those parts of the plant, I put myself in a high-profile position within the company. After about 2 years, I got the opportunity to go into corporate R&D and work with productivity with a wider range of plants, giving me the chance to learn a lot more different processes and products.

Summary

Production Management is a great opportunity for a food scientist. You'll get the satisfaction of making the actual food that people from around the world will eat. It will give you the chance to see a wide range of different products and learn a lot about them. Along with the food, there are also the people. You'll be the boss of a lot of people who are making a good living by being one of your employees. You'll have a lot of good memories.

Chapter 20
Product Development

Brian McKim

Product development is a vital piece of any industry, especially the food industry. It drives innovation, advances technology, and encourages growth. With the increasing concern for the safety of the food supply, product development has become an important step in the overall business process to provide consumers with food they feel safe eating. As a result, the product development profession has become an increasingly important and rewarding line of work.

As the name implies, a product developer is responsible for developing new products to be introduced into the marketplace. This chapter will take a new developer or someone interested in the field through the process outlining project types and the steps of the development process. This is a brief synopsis of what to expect if you are interested in becoming a product developer, as well as how to have a successful product development career in the food industry if you're just starting out.

Types of Projects

A product developer in the broad sense of the word is someone working within a cross-functional team in order to bring a product from idea to shelf. Product developers are creative people who have strong technical backgrounds. There are four major types of projects that a developer can find themselves working on:

1. Line Extensions
2. A New Product

B. McKim
Kraft Foods Inc., Glenview, Illinois

R.W. Hartel, C.P. Klawitter (eds.), *Careers in Food Science*,
DOI: 10.1007/978-0-387-77391-9_20, © Springer Science+Business Media, LLC 2008

3. Cost Optimization
4. Quality Improvement

A developer will be expected to work on all these types of projects; a typical developer will be managing multiple project types simultaneously, all in various states of completion.

Line Extensions

Line extensions are where a developer may be asked to either develop a new packaging size, a new flavor, or a product with a new nutritional profile. A project of this type is needed when a company already has existing technology and internal knowledge on how to produce this type of product. Perhaps the consumer is already familiar with the brand, but new flavors are needed to grow the line and guard against competitive infiltration. The most well-known example is Diet Coke®. When Diet Coke® was launched, Coke® was an already recognizable brand and Diet Coke® was a line extension that played on the consumer recognition of Coke® to gain customer trial. This project type is very product development–driven since the majority of the work in order to launch the product occurs in the prototype creation phase. As a result, it usually involves a small team, with a shorter timeline.

A New Product

A product where neither the company nor the consumer currently has experience with is considered a new product. These products are not usually unknown to the consumer but is a more convenient or value-added version of a current product. A good example of this would be frozen pizza. The consumer already knew what pizza was but would either make it from scratch or go to a pizzeria to purchase it. The idea of have a pizza in a few minutes with little preparation was new to the consumer. In addition, the technology in order to manufacture the pizzas was brand new to the company launching it. Developers will find themselves generally working with a very large cross-functional team that is directly related to the project, as well as having to reach out to other developers in the company not directly related to the project for information. A lot of the responsibility to deliver a successful product is spread throughout the team when a developer is assigned to this kind of project. Consequently, since new products are the most complex, they tend to go through many steps that other types of projects will not need, particularly in the front end.

Cost Optimization

A Cost optimization project can be further divided into two kinds: productivity and technology-driven projects. By and large, these kinds of projects are all changes to existing products, the company and customers will already be familiar with, even before launch. As a result, the developer must deliver a comparable product that has no impact on the flavor profile or processing steps, while at the same time maintaining product quality. An example of a common obstacle is, when a developer may have found an ingredient that meets all production needs and has no impact on the flavor profile. The consumer's expectations of quality will already be established and when placed on the label, their perception of quality may be diminished. Since the consumer is already familiar with the product, these projects can be difficult.

The food industry is very dependent on commodity costs; when commodity costs are high, companies are forced to find cheaper sources. A developer may be asked to explore and implement these alternate ingredients in order to offset the high prices. This type of project would be considered a productivity project.

With technology growing at an all-time high, and new innovative, economical ingredients being brought to the market almost daily, it is the responsibility of the developer to stay abreast of these changes. A developer uses this information to create project ideas and implement them where applicable; as a result, these are technology-driven projects.

These projects are used to keep competitiveness with the pricing of competitors, counteract inflation, as well as to enhance the bottom line in order to fund future projects. All of these are very important for a company's growth—without this project type, a company would solely be dependent on increasing the volume of existing products. They typically involve very small cross-functional teams with nearly all the development driven by the product developer.

Quality Improvement

There is a growing consumer expectation when it comes to food quality. Quality improvement projects fit into one of two major types: first would be where a company has a product on the market but they intend to be better than current, and second would be when a company has a product on the market but they intend to launch a product better than the competition. This type of project has the advantage of having existing products out there to use as a reference point. As a result, it also has the drawback of having to create a novel product that is more desirable over the current product or the competition. These projects frequently involve very large cross-functional teams, with a lot of the responsibility spread among the functions.

Project Phases

No matter to what type of project a developer is assigned, there are always three major phases of product development:

1. Idea Formation
2. Prototype Creation
3. Commercialization

Each phase involves numerous steps, but no matter even if your project is a line extension, a new product, a cost optimization, or quality improvement, your project will go through these phases.

Idea Formation

Many of the new ideas for line extensions, new products, and quality improvements are created by the marketing team; however, ideas can come from anywhere—perhaps most importantly from the development community. This gives a developer a chance to be very creative and have an active role in their company's growth through new product offerings. Just as important as idea creation is idea selection. A product developer may be involved in this step as well and can have significant input in order to assess the technical feasibly of a proposed idea.

Prototype Creation

Initial Preparation

The majority of a developer's job is supporting one of the most important phases of any project—prototype creation. A prototype is a small scale example of what the finished product will look and taste like, and is used to keep development costs down. There are many substeps to prototype creation that a developer needs to consider. In order to have a successful project, a developer will need to do as much research work up front as he/she can. A few examples of some questions that may need to have answered are:

- Is this product expected to meet any type of health claim?
- What are the desired raw material costs expected for this type of product?
- Who is the target consumer? Is the product for children or adults?

A developer then needs to consider how and where the product is going to be made. For some projects, such as line extensions, existing equipment will most likely be used; however, for new products, a process engineer or other technical expert may be needed to help determine the equipment and process. It is very common for some pieces of information to be undefined early on in the prototype creation process. At this stage, the more information a developer can collect the easier will be the subsequent steps of prototype creation.

Raw Material Procurement

Companies will have an extensive library of raw materials that a developer has access to; however, sometimes new raw materials are needed, as in the case of a cost-optimization project. Meeting with suppliers and ordering ingredient samples is another role of a developer. Understanding ingredient specifications and knowing what to order is critical to successful product development.

Working closely with suppliers can be a rewarding relationship for a developer. Suppliers are a wealth of information and a developer can use them anywhere from troubleshooting a problem with a raw material all the way to understanding emerging trends for new product ideas. Suppliers may even go as far as fully developing a product and will just hand it over to a developer to make the product fit within the standards the company has instituted. Since a supplier is developing the raw materials, they will have the best understanding of how the product will work in a proposed system.

Another group a developer should make sure they work closely with and bring in on the project as soon as possible is the regulatory team. There are many ingredients that are not approved for use in the United Stated and it is important for a developer to know this. Certain ingredients and their concentration in products are strictly regulated depending on the country and even state the product will be sold in.

Prototyping

Once everything is ready, a developer can start physically creating the initial prototypes for assessment. Prototypes must be developed for everything from an innovative new product to testing current products with a new minor raw material. No matter what type of project a developer is working on, prototypes must be created. Most of this process is done in a pilot plant or on the bench-top in the lab. A pilot plant is a scaled-down version of the larger plant where the products are intended to be made, whereas bench-top refers to even smaller equipment that may not even resemble what might be used in the manufacturing plant. At this phase, a developer will be producing batches that are a fraction of the batch sizes you would find when actually scaled up to plant production. This saves both money and the time of the developers since they make many more examples in the same amount of time. Once the prototypes are completed, a developer, with the help of the team, assesses the prototypes and gives feedback to aid in moving forward to a plant trial.

During this phase, a developer should work with a trained sensory panel as well in order to get a technical opinion on differences or other specific attributes. If you are trying to match a certain flavor profile for a productivity project, this would be a perfect use for a sensory panel. They are trained to pick up on differences and can give vital insight into what the consumer may pick out as differences during the consumer testing. A sensory panel is typically not used to tell a developer if the product tastes good or if there is a preference over the competition. There are many types of sensory tests that a developer can utilize and he/she will need to work with the panel to choose what is appropriate for their product.

Another group to help facilitate prototype creation may be the statistics group. They can help with initial prototype creation in order to give the cross-functional team options during team evaluations as well as during consumer testing. Most people work better when they are given options and can compare to something else versus when they are asked to describe what needs to change. A common solution to this is called a statistical design. This is where known key attributes of a product are set at varying levels to determine the correct level in the final product. By working with a statistician to generate a prototype design, a developer will be left with a plan with varying levels of each ingredient. This ensures the generation of a set of very diverse prototypes. This test contains some prototypes that may appear to have too high a level of a certain ingredient but since the levels of each variable goes from very low to very high, the correct levels of each ingredient will be obvious and what blend is preferred. One negative to this kind of prototype design is that every time a new attribute is added, the number of prototypes increases exponentially, resulting in a complex prototypes creation phase. It also results in a very expensive consumer test.

Plant Trials

To initiate a plant trial, a developer is expected to submit a plant trial request to the manufacturing facility to get line time, procure raw materials, and verify availability of needed plant personnel prior to the requested date. The request outlines what a developer is anticipating to accomplish, and also contains any other information that is special to the trial such as a sampling plan, what development personnel will be on hand, or what the disposition of the final product will be. A developer will be required to draft a set of analytical specifications to verify consistency of the trial and will need to prepare a HACCP plan.

The food industry prides itself on consistency, both for the consumer's benefit and the company making the product. The consumer expects to go to the store, purchase a product, bring it home, use it, and have the same result as the last time they purchased that product. As a result, a company depends on consistency to maintain preparation instructions, meet consumer expectations, and keep consumers safe while keeping costs predictable.

Drafting a HACCP plan is collaborative effort of the developer, the microbiology department, and the quality department. For most types of projects, a plan is already in place and only minor change forms are needed, but knowing how to do a full HACCP plan is training that is very beneficial to a developer. The earlier in the prototype creation process a developer can begin communication with the microbiology team the better. This is to determine what steps need to be taken to produce a safe, wholesome product and the developer can implement these during the prototype creation phase as well as during the plant trial.

Final Team Assessment and Consumer Feedback

Most products will have gone through many generations of prototype creation and team assessment before the team agrees on a set of prototypes that can be sent into a plant trial. At this point, the final set of prototypes is ready to go to a consumer test. Plant trial product is used for consumer testing since it is exactly what the consumer will see once the product is on the shelf.

There are many types of consumer tests. They can range from a very small, 20-people central location test to a very large in-home use test that can involve hundreds of people. The type of project will dictate how the test is carried out. For a quality improvement project, there might be two hurdles—(1) to be better than current, or (2) to be better than the competition. A developer will work closely with a consumer insights team to help determine what types of tests are warranted for their product. Depending on the results of a consumer test, a developer may need to make changes to the products they have created and retest. Optimally, a developer would have all the information that they need in order to make the appropriate changes to the product as a result of these tests.

Commercialization

Documentation

During this phase, the developer is primarily responsible for bringing together a lot of information and organizing it for the initial plant startup.

A plant startup is a lot like a plant trial and a lot of the same paperwork is needed. The only difference is the plant startup paperwork needs to contain only the raw materials and specifications for the winning prototype.

Application Testing

It may be the responsibility of the developer during this step to understand other feasible uses for the product they have developed and verify that it can withstand these foreseeable applications. The developer may not carry out the testing but will be managing the testing and the information that comes out of it. For example, a piece of cheese can be consumed as is, used in a sandwich, melted over nachos, or used as an ingredient. A test must be created to evaluate these applications and a set of standards established by which the new product will be measured.

Shelf Life/Storage Conditions

For the same reason that it is used for consumer testing, product from the plant trial is also used for shelf-life testing and determining storage conditions. Using plant trial products will give a developer a good indication of what the shelf life and storage conditions are going to be once the product is in the market.

Shelf-life testing involves subjecting the product to conditions that are indicative of what a consumer may do and how the product may taste over time. Expiration dates on products are set using this testing. A developer is trying to determine what effects time and temperature have on product quality and what changes the product may go through. Changes can include separation, discoloration, off flavors, or moisture changes. Shelf-life testing basically measures how long it takes for the product to go from optimal to outside the predetermined standard for product quality.

Storage conditions are very important to both product quality and microbial stability. A product may be shelf stable until it is opened, after which it may need to be refrigerated to retard microbial growth. Understanding consumer usage behaviors has a bearing on where the product will be shelved as well as stored once opening. There is a synergy between these two attributes that can create a high quality, reliable product no matter what stage of shelf life.

Distribution Testing

Distribution tests are performed to make sure that the prototype products can withstand the distribution cycle and what the consumer sees on the shelf is the same as what is seen leaving the plant. Distribution tests can be as simple as internally using a shake table over a few days, all the way up to a real truck taking the products from one coast to the other. Depending on time of year, time of day, and region of the country, conditions can be very different. Just as many products are distributed both internationally and nationally, these products may have to go over the Rocky Mountains in a truck or to the middle of Africa in the middle of summer.

First Production Approval

The whole cross-functional team as well as the plant go over the first production product and discuss any issues that arose out of the first production. Here, at this meeting, everyone tastes the products one last time before they are released from the plant and go out into the market. Also during this meeting, the whole package from case all the way down to code date are checked to make sure for the last time that what is being sent out to the market is what everyone wants to present.

Follow-Up

Over the first six months of production, a developer must keep in close contact with the plant to make sure that no issues arise while running the new product. Since this is probably the first time the product will be produced for hours and in some cases, days on end, there can be new things that may not be evident during plant trial or the startup. This is the last time for the developer to check off and close any open loops before handing it off to the plant.

Important Skills

There are a few characteristics of any new successful product developer that do not fall into a development phase or project type but are important for all phases of development and project types. They are:

1. Teamwork
2. Organization
3. Customer Focus

If a developer keeps all three in mind throughout the development process, he/she will have a successful product development career.

Teamwork

The first skill is to being able to work well within a team, since most companies are simplistically just big teams. The best teams are those with good communication between the various team members. A product developer deals closely with many functions such as marketing, operations, microbiology, engineering, plant personnel, and/or basic research, in addition to other developers. Much of a developer's day is spent collecting information with the help of these other functions most of which are not directly related to the project. Communication with the other functions is very important to a successful product launch as well as a successful career. In most cases, at least a portion of their assigned project has already been done to some extent at some time, and there is either a person or information out there on how to do it. Communicating with associates to find this information is key. It is important throughout the process that there is not inadequate or too complex communication between team members that can slow down the process. Most problems that arise while working in a team could have been avoided with better communication between its members.

For some people working within a team comes very natural, but for others working in groups and depending on other people can be difficult. One good way for an aspiring product developer to become a better teammate is as simple as working in teams, and soliciting feedback to help you better understand where you need to improve. This can be during a chemistry lab or using other classmates to help you study.

Organization

Second is organization; a developer can have a very complex new product project that lasts years, all the way down to many smaller projects at various states of completion that may only last a few weeks. As mentioned previously, a developer needs to reach out to various cross-functional teams, so being organized is very important for a productive meeting. A developer will be responsible for what they have developed, and passing on the knowledge to other developers is critical to a company's long-term success. Being organized is one major way for a developer to set themselves up for not only success today but also in the future.

This is also true for successful students. They must keep good notes, be able to not only manage large projects but many smaller assignments as well, and have good time management skills in order to coordinate everything together. Students who work on these traits will not only ensure a successful professional career someday but a successful academic career today. Three easy things everyone can do are (1) take the time to review your notes, clarifying any vague points, and preferably the same day, (2) create a system to keep updated on the current status of assignments and any next actions that need to be taken to move them forward, and (3) prioritize the items on your to-do list.

Customer Focus

Lastly, it is critical for a developer to understand and appreciate who they are developing food for—the consumer. Understanding the consumer and what he/she expects to get out of a product is a valuable attribute. Most consumers will go through a cycle of purchasing a product, assessing the product, and then deciding if they will repurchase it. Identifying what the consumer liked and disliked about the product that causes them to either not purchase or repurchase is a great place for a developer to begin idea creation for many types of projects. Obviously not all consumers have the same desires for a product. Two consumers may buy a product for two very different reasons—one may be purchasing for the nutrition profile and another may be totally price driven. There is a need to recognize these differences as a developer.

As a student getting in tune with the consumer can be difficult. Though there are a few ways they can do this while still in school. Most emerging trends will not hit the mainstream market for a few years. Things that are big in the food industry now were emerging trends three years ago in restaurants. By reading the restaurant section of local and national newspapers or just looking at a restaurants menu, you will begin to see these trends and begin to understand the consumers.

Conclusion

It is important for a new product developer to understand that there is a lot of criticism in this job, albeit mostly constructive criticism. However, coworkers and the consumer may not always like what a developer has produced. It is important to understand that this criticism is not personal and as long as the product developer remembers that he/she is trying to please the consumer (and that they themselves may not always be the target consumer). In addition, projects can be put on hold or even stopped at any stage of the development process. This can be especially true with new products, many of which may never make it to market. Project termination can occur for any number of reasons and it is not the fault of the developer.

There is nothing more rewarding for a developer to walk down the aisle of a grocery store and see his/her products on the shelf and in people's grocery carts. It is a very rewarding position that has very tangible results.

Chapter 21
Technical Sales

Rose Defiel

A career in technical sales, working with a company that supplies ingredients to food manufacturers, can be interesting, challenging, fun, and an opportunity for personal growth. Technical support can be involved in every step of the progression, from when the new project comes in until the customer is buying your product and using it in their process. Depending on the size of your company and how it is set up you could personally be involved in one or all of the areas.

Technical sales is a support function. You will enjoy working with every department within a company. Technical sales is the department within a supplier company that understands not only the products being manufactured, but the processes used to make the products as well as the end uses of the product by the customer. Technical sales is often called on to answer questions from internal (within your own company) and external (the manufacturers who purchase your ingredients) customers. They are also looked to for suggestions on how to improve or develop specific products.

A major portion of time is spent working with the sales department to not only help drive new sales with new or existing products, but to maintain the current customer base by providing information and support to them. Many manufacturers are looking for ways to cut costs within their organizations. This leads to customers (the food manufacturers) pushing product development down onto their suppliers. Customers can expect you to develop new products independently, or they may spend time working together with you.

Technical sales require not only the scientific knowledge of your own product, but excellent people skills, good listening skills, good problem-solving skills, and knowing how to ask questions that will provide answers,

R. Defiel
Clasen's Quality Coatings, Middleton, WI, USA

R.W. Hartel, C.P. Klawitter (eds.), *Careers in Food Science*,
DOI: 10.1007/978-0-387-77391-9_21, © Springer Science+Business Media, LLC 2008

which will help you in your quest. You should also be able to and enjoy traveling. Your customers can be anywhere and to support them and your own sales people, you have to be able to travel to the customers, sometimes with short notice.

A technical sales person has to be up-to-date in their knowledge of their own products and their company. They also need to have a deep understanding of the industry they work in and a strong knowledge of their customer base. Understanding what happens to the product you produce once you sell it, and the potential different applications and uses is essential. A technical sales person must be well versed in the hottest topics and newest trends in not only their product line, but in the industry as well. A technical sales person can be involved with developing cutting edge processes or products.

A technical sales person continually learns and explores for more information. Reading technical journals related to your products as well as your customer's products is important. Attending seminars and training sessions for your product as well as your ingredients is important for staying in the forefront of your industry. Your company wants others in the industry to think of you as an expert when a question comes up or information is needed.

Many technical sales people have an advanced degree, usually at least a master's degree. Most people in technical sales have extensive experience working with the products they produce, or related products. Understanding and knowledge in some areas can be applied to other areas within the food industry; for example, expertise in fats and oils can be valuable in the confectionary industry.

The knowledge and preparation for a career in technical service can be gained in different areas of a company, but a quality or research and development (R&D) background gives a good understanding of the important parameters of the product. Quality and R&D also require a fairly good understanding of the process used to manufacture a product. Generally people working in the Quality or R&D departments have a technical degree that will also be helpful in technical sales.

A job in technical sales will entail being involved in many different functions. Some of the things you will be involved in are writing technical papers, giving technical presentations to customers, attending sales calls, and working with customers on different projects. You could be called on by your own company or your customers to provide troubleshooting help and you most likely will be expected to attend trade shows with your company. These are outlined later to give you an idea of what you can expect in an exciting job in technical sales.

Writing Technical Information and Papers

Customers are constantly asking for information on your products. Technical support personnel write up documents for the sales force to provide to your customers. How to handle the product, what parameters are critical for the functioning of your product, what issues can arise if the product is handled improperly are some examples. These can be in the form of an email, a technical brochure that the sales team passes out, assisting in the design of the company's Web page or standard handouts that can be given out when questions are asked.

Seeing trends in the questions and information requested can help a technical sales person be proactive and have information readily available. Being up to date on issues and trends in the industry is also helpful so that responses can be quickly composed.

Attending Sales Calls

Often when sales people call on customers, they are dealing not only with the procurement professionals but they also meet with the food scientists or product developers employed by the customer. Being present during the sales call can allow you to help answer questions, suggest products, or offer suggestions—something that can be very valuable. Participating in sales calls usually requires traveling to the customers offices.

Providing Technical Presentations

Customers may not understand what your company has to offer. They may not be familiar with your product or how they could use your product. They may not understand the capabilities of your R&D department or of your company. They may want to know what is new in the industry and how it could potentially affect them. You may be asked to give technical presentations to a person or a group of people detailing any or all of the aforementioned topics. Each technical presentation is tailored to the specific customer and their requests and expectations.

Sometimes the request is for a general "tell us about your company" presentation. It could also be detailed as to how your product would fit into their application or focus on the current trends in the industry. Giving a technical presentation requires a firm knowledge of the topic so that any questions that arise during the presentation can be answered. The presentation also should be written for the audience. If you will be speaking to a group of sales people, you would probably write differently than if you were speaking to a group of food scientists. Communication with your own sales team is important in making sure that your presentation is appropriate for the meeting.

Technical presentations are usually computer-generated using Power Point or a similar tool. Computer skills are important to designing an interesting presentation that will keep your audiences attention. Handouts can also be provided with the presentation. The details of what to bring and who is bringing it should also be established with the sales team prior to the meeting.

Project Work

Customers can request new products, new ideas, or modifications to existing products. These are usually submitted as projects by the sales people. Sometimes the request is a only a concept or it could be a request for a product that will do something specific, such as melt in your mouth, or give the final product a certain attribute, like making your mouth tingle when you eat it. Technical support is required for the project and ensures that all of the necessary information is covered. This usually involves talking with the sales people as well as the customer. Understanding the scope of the project, the customer's needs, and timeline is critical for developing a product that works.

The more information that can be obtained upfront the smoother the projects usually go. Developing forms to assist the sales people in asking the pertinent information can be a good first step. Understanding what information is needed to start a project helps the sales people provide better information. Types of ingredients wanted, allergens, functionality, and price are all important considerations.

Once the project starts, samples are made and sent to the customers for evaluation. Often times the project changes over time so you have to be flexible and open to making changes. Projects can change course for different reasons. Cost is usually a big reason; the customer may like the product, but want it to be less expensive. Sometimes, once the customer's sale and marketing people get involved they may decide to go in a different direction. The customer may find that what they originally had in mind for the product does not give the desired result in the final product. The customer may submit the finished product to a taste panel, which can also lead to changes. Your product may not work efficiently or properly in the customer's process. Listening to the feedback provided by the customer is key to staying on course and having a successful project.

If a customer is using your product for the first time, there may be a need for them to visit you, or you to visit them to gain an understanding of the project. They may need guidance in the use of your product. Working with not only the food scientists but the engineers and operations people can be a vital part of the process. You may be providing the engineers with the information they need to design their systems to support your product. You could be explaining to the operations people how to handle your products and what factors are critical in the products working properly for them. The goal is for your company to sell a product that works for your customer, not just develop a product.

Often when your customer uses your product for the first time they request technical support. This is where it is important to understand your customer's final product as well as the production methods they use in their facility. It is important to understand the process so you can help the customer get your product to run.

Troubleshooting

Often when customers have problems running items containing your products they will contact you for assistance. Again, understanding the uses of your products will help you to help the customer to troubleshoot. This usually takes place through a phone call, conference call, or email. Even if the problem turns out not to be your product, but their process or the handling of your product, the ability to offer the customer a solution is very important.

Troubleshooting can be as easy as a phone call. It can also involve getting samples of their product, your product, and pulling "retain samples" for analysis. If the problem is not easily solved it could include a site visit to the customer. Seeing the process for yourself can give you an insight into a problem. The ability to get their process to work can mean keeping your product in their facility (and continued sales).

Trade Shows

Trade shows can be an important opportunity to present your product to existing and potential customers. Trade shows are usually staffed with both sales and marketing and technical sales people. This allows the company to hopefully provide all the information that the customer is looking for in one stop. It is a valuable tool for your company to reach many different customers in a relatively short amount of time.

Trade shows are a great place to interact with both suppliers and customers. Your company and products get exposure to both current customers and potential new customers. People within the industry go to trade shows to visit suppliers, look for new ideas, and see what is new in the industry. Companies set up booths to display their company and their products.

People will visit booths that look interesting or that have an ingredient or process that might benefit their product. Technical sales people work at trade shows so that when customers want to discuss issues or ask questions, the support is right there. Technical sales can also make recommendations for specific products that may meet a customer's need. Having the expertise onsite at the show helps to portray the company in the best light and provide immediate assistance to the customers.

Summary

These are examples of the types of interaction you can have as a technical sales person in the food industry. Travel can be a big part of the job, so you have to be able and willing to travel. The opportunity to meet new people, be involved with cutting-edge technology, and use your food science training are all reasons to consider technical sales. Technical sales is a challenging job that allows you to work together with many different people within an organization. In order to participate in the many different functions that you will be called on to support, you will need to have a deep, solid understanding of your product, its uses, and your customer's applications.

Customer support through project work, troubleshooting, or providing presentations to them can be important functions. The way you perform these duties can make the difference between gaining a new customer or retaining one you already have. Customers like to work with companies that are able to support the product they sell. Having a resident expert puts a company one step ahead of the competition.

Chapter 22
Day to Day Life in Research and Development

Dennis Lonergan

One good way to figure out if a career might fit your interests is to take a look at what the day to day activities would include. This is often less glamorous than a generalized view, but helpful nonetheless. For instance, being a dive master in the Caribbean might sound very glamorous, but the day to day activities, such as lugging around customers' heavy gear and dealing with demanding vacationing clients is also a part of the job. So let's take a look at the day to day activities of an research and development (R&D) food scientist and some of the key variables that will affect those activities.

In general, an R&D food scientist will be involved with doing something "new." This is often new to the company rather than new to the world. For example, if you worked for a meat processing company, developing meat-filled, refrigerated pasta items may be new to the company, but it would not be new to the industry. Similarly, you may develop a slow flavor release product that was new to the food industry, but it applied technology that was well known in the pharmaceutical industry.

There are probably many ways to differentiate "research" from "development." For this look at day to day activities, I'll define "development" as taking what is known, and applying it or putting it together in a new way. This may be novel to the industry or novel to your employer. Either way, it is new and different, and involves change. And change is always a challenge for people individually and for organizations.

"Research," by contrast, will be defined here as doing something that involves building totally new knowledge. Often it is a matter of the degree of the challenge that differentiates if the work is described as research or development. For example, if your project objective is to develop a soup that tastes the same as the current version with 75% less sodium, that would probably

---

D. Lonergan
The Sholl Group II, Inc., Eden Prairie, MN, USA

R.W. Hartel, C.P. Klawitter (eds.), *Careers in Food Science*,
DOI: 10.1007/978-0-387-77391-9_22,

involve new technology and thus would be an example of "research." If the goal was a 20% reduction in the sodium content without changing flavor, the tools to do this may be known, and then it would be a "development" project.

There may also be a time when a project transitions from "R" to "D" (or it may be called transitioning from R&D to product development). Projects, be they research or development, start with a desire to do something "new." This desire may come from marketing, sales, operations, or R&D/product development itself. You can then ask the question "is this something that we have the tools and knowledge to do?" If the answer is "yes," then it is a development project. If the answer is "no," then it is a research project.

An example could be a desire to have a croissant with zero saturated fat, and it tastes every bit as good as a 28% butter croissant from the local bakery. This is not something that anyone knows how to do, so it would require significant research. In this case, it would probably require both new product and process technology. After the research develops the tools and knowledge to enable the "new" thing to be developed, it then becomes a development project. In large companies, and especially in other areas such as the pharmaceutical and medical device industries, these R&D groups can be very separate functions. In other cases, the transition from research to development can be gradual and possibly even within the same group.

Type and Size of Company

Consumer package goods companies are often the first to come to mind, especially the large corporations such as Nestle, Kraft, General Mills, Kellogg, or Con Agra to name a few. Food scientist's also have R&D jobs at many smaller food companies. The reason I bring it up here is that the size of the company will affect your day to day job.

In general, the larger the company the more specialized your job will be. As you look at the list of day to day activities that follows, in a large company you may spend any given day doing one item on the list, and may well even be part of a group of people doing that task. In a very small company, you may work on many of the activities, and may do so without much assistance from others. You will need to discover for yourselves which suits you best.

In addition to consumer products companies, there are many other firms involved in supplying food to our population, all of which need R&D food scientists. These include, but are not limited to:

- producers,
- ingredient suppliers,
- packaging material and equipment firms,

- retailers (e.g., development of private label brands), and
- companies that just do R&D for other firms.

As with the more familiar retail consumer foods companies, there are large, small, and everything in between corporations in terms of size. A good way to get an idea for the breadth of companies is to look at some trade journals at your local college library. Some examples include *Food Technology, Food Products Design, Prepared Foods,* and *Food Business News*. Many of these publications have an annual buyer's guide, which will show you the breadth and number of firms in the food industry. You may also find versions of these publications on the Internet.

Types of Projects

Developing new products is probably the first thing that comes to mind when thinking of the types of projects that an R&D food scientist will work on day to day. These can be new products that the consumer sees directly, or ingredients or processes that enable retail food companies to make new products. "New" products can range from "line extensions," such as a new flavor variety, to "new to the industry" offerings, such as self-rising pizza crusts. New may involve a different final product, or it may be "new" in that it adds more convenience without changing the product per se. The introduction of individual pieces of refrigerated cookie dough, rather than having the consumer slice a piece of dough from a cub package, is a good example.

These activities certainly require the skills of a food scientist, both in the area of product and process development. However, there are other types of projects that require a food scientist that are just as vital to the success of a food company. These include ingredient and process improvements that improve the flavor, nutritional quality, or reduce the cost of production. An example is reformulating a product to accommodate changing availability or cost of ingredients. Another would be reformulating to eliminate trans fat from a product.

Day to Day Activities

The following is a list of a number of activities that provide the "how" part to the types of projects listed above. A sports analogy is that winning in baseball is about scoring more points than your opponent. The "how" gets into pitching, hitting, running, catching, etc. These will be key activities in your job if it is either "research" or "development."

- *Reading and speaking with others.* This is critical in being a good R&D food scientist. There is a wealth of information in the literature, both academic and from that published by suppliers. I cannot overrate the importance of using it to your advantage. Also, asking coworkers for their ideas and experiences is a good idea. "What do you know about...," should be a well used phrase in your vocabulary.
- *Bench-top formulation work.* This is where many ideas are born or killed. An example would be trying out the idea of making low-sodium bread by just removing all of the salt from a bread formulation. The results will probably suggest that there was a reason why salt was there in the first place!
- *Pilot plant development work.* The pilot plant can be just as valuable as the lab for generating and testing new ideas. It contains equipment that is larger that what you would find in a kitchen or laboratory, but smaller than that found in the production plant. It is also the place you go when someone says "that looks great, but how are you going to make it?" Both the pilot plant and the bench-top laboratory are areas in which a food scientist may spend the majority of their time.
- *Sensory analysis.* This is the technical way of answering the question "how does it taste." It is a very important area of food science. The tests that are used can be grouped into the types of questions that are to be answered. This is a good point to mention again the difference between small and large companies. In a large company, you probably will give your samples to someone else to test. In a small company, you may well be going down the hallway asking people to taste your samples and then analyzing the data yourself.

  • *Difference Tests.* There are several kinds of difference tests, one of the simplest resembling the kid's game of under which of three shells is the pea hidden. In this test, a person is given three samples to taste. Two are the same and one is different. They are then asked to identify the one that is different than the other two. You then do this with about 30 or more people, and apply statistical analysis to the results. This test is often used to answer questions like "does the product taste the same if I replace ingredient X with ingredient Y" or "if I remove 20% of the sodium, can people tell the difference."
  • *Attribute Tests.* This type of test is often done with trained tasters. Attributes to be tested can be as straightforward as sweetness, saltiness, or crispness. It can be used to find out "what is different about these two products." In it, people rate the intensity of a certain flavor, texture, or "attribute." Again, the results are statistically analyzed.
  • *Consumer preference tests.* If you want to find out how much people like your product, then you need to go out into the real world and have

them taste your product. It's not fair using your coworkers or friends. This can be done with the people coming to a central location and tasting product, which has been prepared for them, or by having them take the product home and preparing it themselves. The latter test also answers ease of preparation questions.

- *Listening to consumers first hand.* The "focus group" method entails consumers gathering in a room with a moderator. There is usually a one-way mirror at one end of the room. You will be behind this mirror. This format can be used to gather consumer reaction to a new product concept, or to engage them in a conversation about a type of product or meal occasion. A more real life method is to go into a consumer's home and actually film them as they go about using your product (or a competitor's). This is called an ethnography study, and it can be very informative (and humbling) to see a consumer struggle with your "easy open" feature and "simple preparation" directions.
- *Package design and labeling.* Package design is critical in that it can have a huge impact on the consumer's decision to purchase your product. It also impacts their use of your product, as well as conveying important, and legally required information. Ever wonder who wrote the directions, ingredient statement, or put together that complicated nutritional information panel? It was probably a food scientist.
- *Shelf-life studies.* Part of the responsibility of the R&D food scientist is to make certain that the product not only tastes great when the company puts it in the package, but that it still tastes great when someone consumes it some time later. How much later? It could be a matter of days for things like fresh produce, or up to a year for canned goods. In addition to protecting the product, you want the package to be both cost-effective and environmentally friendly.
- *Recording your results.* Just like the chemistry or biology lab that you had, which required lab write-ups, you will need to record what you did and the results that you obtained. This is not only necessary for you keeping track of your own work, but also so that others can learn from your efforts. It is also critical for "intellectual property rights" (patents).

How to Prepare for an R&D Job: What to Do While You Are Still in College

- *Prepare for graduate school.* As an undergraduate, this may not be what you were looking forward to hear. But before you skip to the next section, let me explain why you should think about this possibility.

- if you want to work in R&D for one of the large food companies, they may not even talk to you unless you have an advanced degree. For smaller companies, that may or may not be the case. But, are you certain at this point in your career that you want to close off the possibility of working for a large multinational food company in R&D?
- I've never meet anyone yet who said "I wish that I hadn't gone on to graduate school." I have meet many who said "I wish I had," or defensively said "I could have."

So if you are still with me, what might you do to prepare for graduate school?

- Get good grades. Most graduate schools require a 3.0/4.0 GPA as a minimum, and the more competitive schools are just that, more competitive to get into.
- Get a good sound technical background. An extra course or two in chemistry or biology may be useful, and hopefully interesting, as you do see yourself as a scientist working in R&D, don't you? Taking the "harder" version of a science series may also be a challenge that you end up being proud to have undertaken.
- Feed your curiosity. Read about current technical topics in food science. What will food and food production look like the future? Why all of the interest in probiotics? What role do you see yourself and the food industry playing in the challenge of obesity? Is food bioterrorism a concern?

– *Networking*. Join your school's food science club and get involved. There should also be local sections of Institute of Food Technologists (IFT) and the American Association of Cereal Chemists International (AACC). These groups usually welcome students and often offer a good meal at a reduced rate. I always hated the saying "it isn't what you know, but who you know" as I thought it belittled the value of education. However, I now have to admit that meeting people, smoothly impressing them with what you have learned and with who you are will be helpful for your career.
– *Semester abroad*. As with getting an advanced degree, I have never met anyone who said; "I wish I hadn't spent that semester abroad." There are a lot of opportunities available, and most colleges have programs to make it easier and have set up procedures to expedite things like making certain credits will transfer. So how is this going to enhance your career in R&D? Many of the large food companies in the United States are now multinational corporations looking to international markets for future growth. Thus, experience abroad can be one more thing that sets your résumé apart

from the crowd. A less pragmatic reason is that experiencing a new culture will open your mind to different ways of looking at things. Many of these experiences won't specifically deal with food science, but hopefully it will help you bring an open mind to your R&D problem solving down the road.

– *Internships*. This is an excellent way to find out if you really want that job in R&D. The money is typically better than most summer jobs, and it is a great way to get going on that networking. For most companies, the summer intern program is part of their overall recruiting effort, so that summer internship may lead to a job offer in R&D.

Summary

I hope that this sounds like the type of day that would keep you wanting to come back to work the next day. If so, you may have found the major and career that is right for you.

Chapter 23
Government Regulatory

Katie Becker

Government regulation of food products, food processing, and food preparation is imperative in bringing an unadulterated, nonmisleading, and safe food product to market and is relevant to all areas of food science, including engineering, processing, chemistry, and microbiology. The liability associated with providing consumers with an adulterated or substandard product cannot only tarnish a company's name and reputation, but also impose substantial financial repercussions on the company and those individuals who play an active role in the violation. In order for a company to fully comply with the relevant food laws (both federal and state), an intimate knowledge of food science is required. Individuals knowledgeable in food science play an integral role not only in implementing and counseling food companies/processors to ensure compliance with government regulations, but these individuals are also necessary to the state and federal governments that make and enforce the relevant laws and regulators. For these reasons and more, to be further explained later, government regulation of food protection and processing presents many diverse career options for a food scientist.

Federal Regulations

Food regulatory law encompasses many areas of study including food science, business, and law. The primary reasons for enacting food laws include prevention of foodborne illness and preventing consumers from receiving illegitimate or adulterated products. The two main government agencies that regulate foods are the Food and Drug Administration (FDA) and the United

---

K. Becker
Attorney at Law, Chicago, Illinois

R.W. Hartel, C.P. Klawitter (eds.), *Careers in Food Science*,
DOI: 10.1007/978-0-387-77391-9_23,

States Department of Agriculture (USDA). The FDA is responsible for enacting and enforcing laws including, but not limited to, the labeling of foods, setting standards of identity for food products, and approving and regulating food additives and GRAS (Generally Recognized As Safe) substances. The FDA also enforces and regulates its laws in the following areas: misleading products, mislabeling of products, the contents of a food label, and nutrition facts and claims (i.e., health claims, nutrition claims, and qualified health claims). The two major roles the FDA plays are to administer inspections of food plants (and thereby protect the health of the public) and to test and set standards for products. Furthermore, the counterpart of the FDA, the USDA, concentrates its regulatory efforts on compliance in the meat and poultry industry.

With innovation at its height, as many new ingredients, additives, and new technologies (e.g., implementing nanoscience into foods and food processing) are being developed daily, food regulatory requirements are at elevated levels of importance. Labeling issues, evaluation of GRAS status for packaging components and food ingredients, developing and implementing FDA compliance procedures and implementing responses to government inspections (i.e., recalls) are taking a front seat following threats of bioterrorism in the food supply, highly publicized and nationalized foodborne disease outbreaks, and new food products and ingredients being developed at rapid-fire pace.

One technology in particular, application of nanotechnology in foods, will pose many challenges for the FDA and will require the knowledge and expertise of food scientists to assist the government in regulating this emerging technology. In particular, areas of interest in food nanotechnology include: nanoparticles in edible coatings and barriers, preservatives, antimicrobials, and mineral supplements. With the possibility and probability of applying nanoscience in food packaging and processing and ingredient technology, interest in this emerging technology is especially prevalent to food companies and is also resulting in increased private funding in this area. In order for food nanotechnology to be approved, accepted, and implemented into food-related applications, the involvement of food scientists in obtaining the requisite government approval is necessary.

The most prominent statute enacted and enforced by the FDA is the Food, Drug, and Cosmetics Act (FDCA). The FDCA is a strict liability statute, which imposes criminal penalties, seizures, and injunctions on individuals or corporations who violate the FDCA. Products also may be recalled either voluntarily (by the company) or by an order from the judge (court order).

In addition to lawmakers, attorneys, and lobbyists, scientists also play a pivotal role in the government/regulatory realm. Job opportunities for food scientists in the government/regulatory arena incorporate many facets of the field of food science, and include, but are not limited to:

- food analysis (studying the biological effects of various agents commonly found in foods, such as additives or contaminants);
- food chemistry (conducting research projects that study the effects of food components and dietary supplements on utilizing essential and toxic minerals in the diet);
- food process engineering (presenting reviews, conclusions, opinions, and recommendations to appropriate scientific review panel on premarket approval applications, product development protocols, and petitions for reclassification.); and
- food microbiology (conducting research on the development of media and procedures for isolating and identifying pathogens from foods and on the definition of the kinetics of growth, survival, or destruction of food-borne pathogens under the environmental conditions occurring during food processing and storage).

Even though the FDA is headquartered in Washington, D.C., it has district offices scattered throughout the nation, including Chicago, Dallas, Baltimore, and Minneapolis.

If an individual is interested in not only the laws themselves, but the science behind the laws, a career in food regulation will likely be a suitable fit. Analyzing, interpreting, and implementing laws are also crucial in the food regulatory arena. Therefore, if an individual is interested in food science (processing, engineering, chemistry, microbiology, etc.), but would like to use their food science knowledge in contributing to and/or analyzing and implementing food laws and explore a career outside of the well-recognized food science careers (i.e., research and development (R&D) and quality control/assurance) the career and internship opportunities in food regulation should be considered.

As mentioned previously, the USDA regulates meat and poultry products and processing. Opportunities for food scientists in the USDA include: meat and poultry plant inspectors, food microbiologists, and the like. Similar to the FDA, the USDA also creates and enforces laws and regulations in the meat and poultry industry, with respect to labeling, packaging materials, additives (traditional additives in addition to radiation used to reduce microorganisms in meat and poultry products), and allergens, in addition to performing safety inspections of facilities. Inspectors for the USDA must be knowledgeable in food science applications such as food processing, engineering, and microbiology to ensure that meat and poultry facilities are functioning in conjunction with the standards set forth by the USDA.

Although the USDA is best known for its regulation of meat and poultry products and processing, it has carved out a niche for its technology and intellectual property management (for example, patenting new and emerging technologies in the meat and poultry industry). The USDA partners with commercial firms to transfer its technology to American farmers, businesses, and consumers. The USDA offers private sector businesses, state and local governments, and universities the opportunity to license federally owned inventions. In the words of the USDA, these partnerships are designed to "expedite research results to the private sector, exchange information and knowledge, stimulate new business and economic development, enhance trade, preserve the environment, and improve the quality of life for all Americans." Some patent applications and issued patents available for licensing through the USDA are as follows: "Sweet-N-Up" A new Distinct Peach Variety; Gene That Extends Fruit Shelf-Life; New Technique to Eliminate Bitter Compounds in Potatoes; and New Edible Food Coatings. The USDA is also responsible for the National Organic Program and Organic Foods Production Act, for certifying foods as organic, "to assure consumers that the organic foods they purchase are produced, processed, and certified to be consistent with national organic standards."

Many subsections and specialties exist within the umbrella of food law. For example, some practitioners specialize in packaging law, compliance, GRAS approval petitions and litigation surrounding violations and/or foodborne disease outbreaks. Packaging law relates to the regulation surrounding both the packaging and labeling of food products in conformity with FDA regulations, whereas compliance refers to counseling food manufacturers and processors to ensure compliance with the relevant foods laws. Additionally, GRAS approval petitions require not only legal counseling, but also counseling by a food scientist in order to perform the relevant testing and research and opine as to the safety of a substance in a food product in order to obtain government approval for using the substance in a food product. The knowledge and experience of a food scientist, in litigation surrounding violations and/or foodborne disease outbreaks is also necessary, as food scientists are used as expert witnesses and are needed to build both sides of the case.

State Regulation

The states also play an important role in regulating food products and the food industry. For example, in Wisconsin, the two enforcement bodies of the state government are the Department of Health and Human Services along with the Department of Agriculture, Trade, and Consumer Protection.

It should be noted that the state laws that regulate in the same area as federal laws cannot be more lenient than the existing federal law; however, they may impose stricter guidelines. Additionally, states have embargo-type power, which allows them to halt a product's movement in interstate commerce; however, the FDA does not have the power to go into a food plant and seize/embargo it. Yet, the FDA can take action against anyone in the chain of the product's movement, including production, distribution, and retail.

In recent years, there has been a push on the part of the FDA to streamline states' regulations. As stated in the September 2007 issue of the *Journal of Food Technology*, in order to achieve consistency throughout the states, the FDA is urging states to adopt the Manufactured Food Regulatory Program Standards "for measuring and improving the performance of state programs for regulating manufactured food and help the state and federal authorities reduce foodborne illness hazards in food facilities." These Standards define best practices for the critical elements of state regulatory programs, and include: staff training, inspection, quality assurance, incident investigation, and enforcement, etc.

In addition to federal and state regulations, if products are marketed abroad, they are also subject to international regulations. International regulatory groups include the Food and Agriculture Organization and the Codex Alimentarius Commission.

My Experiences

Due to the variety of disciplines encompassed in food regulation, this career path sparked my interest. As an undergraduate majoring in food science, I was particularly interested in the laws surrounding the processing, distribution, and sale of food products. Following graduation with a BS in Food Science, I interviewed with and obtained an internship through the Wisconsin Department of Health and Family Services (DHFS) in the Food Safety and Recreational Licensing Division.

During my time at DHFS, I evaluated the efficacy of the Wisconsin Food Manager Certification Program, a program included in the Wisconsin Food Code and enforced by DHFS. This program requires that at least one certified food manager is employed in the particular eating establishment, and is based on the establishment's size and/or type of food being served. This research was supported by a grant from the Centers for Disease Control (CDC).

In evaluating the Wisconsin program, I met with interested parties, such as the Wisconsin Restaurant Association to obtain their feedback and opinions on the program. Additionally, I researched and contacted other states with certification programs and attempted to correlate the type of certification program implemented to the number of instances of foodborne illness complaints in that state during a specified period. I also met with representatives from the Department of Agriculture, Trade, and Consumer Protection (DATCP) and the Environmental Protection Division of DHFS regarding the Food Manager Certification Program. An additional duty of my position entailed digitizing a database including policies passed by the Division.

In addition to internships with state regulatory agencies, an undergraduate can gain experience in the governmental regulatory arena by taking food law courses, and interning with a food manufacturer or processor. Internships with the FDA and USDA (in both the national headquarters and district offices) are also ways for an undergraduate to gain invaluable experience in this field.

After working at DHFS for a year, I attended law school, planning to specialize in food law and/or intellectual property law (in food science and the chemical arts). During my undergraduate studies, I took a food law course, which sparked my interest in the regulatory arena of food law and also took a food and drug law course during law school. However, although many law schools do not offer extensive food and drug law electives, the food-related agencies, such as the FDA, are discussed in a variety of courses offered by law schools, such as legislative process and administrative law. Food science–related issues are raised in many intellectual property law classes, such as trademark law, trade secret law, and patent law.

During law school, I clerked at one of the largest food and beverage corporations. In clerking for this corporation, I experienced first-hand how the FDA regulations governed many aspects of the legal department and the corporation as a whole. Any food corporation needs to keep abreast of any labeling laws and all other pertinent regulations, to ensure compliance with these laws. The federal laws that are most integral to most food corporations include labeling, including ingredient labels and claims, in addition to standards of identity, certifications, and the like.

While in law school, I found that my food science background was an invaluable asset to my legal education. The technical writing required in many of my food science courses helped me to seamlessly transition into legal writing. In addition, the time spent researching in preparation to write technical papers and perform experiments and independent study projects also proved advantageous in helping me excel at legal research. Additionally, a technical science background, such as food science is required in order to sit for the patent examination to practice before the United States Patent and Trademark Office.

Summary

Many diverse and exciting opportunities are available for food scientists in the government regulatory arena. Opportunities arise not only in the federal government, but also in state governments and private food companies. The federal and state governments conduct the research behind, implement, and enforce the laws, whereas an industry must ensure its compliance with these laws. Whether your interest lies in bacteriology, chemistry, engineering, processing, etc., the state and federal governments in addition to private food companies provide a wide and interesting array of career options for the food scientists.

Chapter 24
Using Food Science in Special Interest Groups

Alison Bodor

Employment opportunities are excitingly broad and varied for food scientists. There are many special interest organizations including food trade associations, commodity promotion groups, and consumer advocacy organizations that require the skills of a food scientist. Unique aspects of these employers and jobs will be explained along with the special food science and related talents that contribute to success in these fields. I have had the good fortune of using my food science background working for a trade association in Washington, D.C. and I will use my own job as an example. After all, I work for a wonderful sector of the food industry—the candy industry.

What Are Special Interest Groups and How Are They Unique?

An interest group (also called an association, advocacy group, lobbying group) is an organization whose purpose is to advocate for a cause, an industry, or a demographic sector. Interest groups related to food are often established to promote and protect a sector of the food industry. A trade association is an organization made up of business competitors. Businesses—not individuals—join trade associations. For instance, the National Confectioners Association (NCA) is made up of candy manufacturers across the country and internationally. Although these companies are competitors, they rely on NCA for information about legislative and regulatory actions that may affect their business; research and statistics on trends in the industry; professional education; and communications and trade activities to promote the industry.

A. Bodor
National Confectioners Association, Vienna, VA, USA

R.W. Hartel, C.P. Klawitter (eds.), *Careers in Food Science*,
DOI: 10.1007/978-0-387-77391-9_24,

Sectors of the food industry that are represented by interest groups or trade associations include most commodities (e.g., eggs, dairy, milk, corn, meat, poultry, grain, peanuts, almonds, nuts), other ingredients, and processed and specialty foods.

There are also food-related associations that represent a profession such as the Institute of Food Technologists, American Dietetics Association, or the Food and Drug Law Institute. These are professional societies that individuals join (rather than businesses or companies) to learn the most up-to-date information about their profession and share common problems and solutions with others.

Some associations or special interest groups are formed to advocate specifically for consumer issues. In addition, there are many philanthropic associations such as the American Cancer Society, the American Heart Association, and the American Diabetes Association that have programs related to food and food science.

Most food associations are small organizations with less than 50–100 employees. My own association is considered average in size with a staff of 25. As a senior member of this staff, I have a lot of responsibility and am expected to participate in the overall leadership of the association. As with most small organizations, everyone wears a lot of hats and some afternoons I can even be found in an assembly line in our conference room, filling gift bags with candy that will be used to demonstrate the array of our products to other organizations we work with.

Associations are membership organizations. Depending on the association, companies or individuals pay annual dues to be a member. A professional staff runs the day to day activities of the organization, but the overarching goals and objectives of the association are generally set by a governing board of members. Members also participate on committees and task forces that are led again by the association staff. Association staff generally work very closely with members.

Most associations are not located near traditional food processing centers. The Washington, D.C. metropolitan area has the highest concentration of associations in the United States. In fact, associations/nonprofits are the third largest industry in the D.C. area, behind the government and tourism. Many associations are in the D.C. area to work with Congress and regulatory agencies to keep abreast of general policy issues. Some food and commodity associations are based in closer proximity to their agricultural base. For example, the Almond Board of California is located in California where their commodity of interest is grown. The Northwest Food Processors Association is located close to their membership base in Portland, OR. The American Institute of Baking is headquartered in Kansas, a region associated with wheat and grain production and research.

What Are the Responsibilities of a Food Scientist Working for a Trade Association?

Every organization and position is unique, so job responsibilities will vary accordingly, depending on the focus of the organization. I can best answer this question by describing my responsibilities as Vice President of Regulatory Affairs at NCA. I am generally responsible for food safety and regulatory issues affecting the candy industry. I work in a team environment with other staff members in charge of communications and legislative affairs to cover other public policy issues relevant to candy such as diet and health issues and international trade policy. A job description for me might include the following requirements:

1. Assure that members are aware of and have the tools to comply with regulations affecting candy.
2. Work independently and with other food industry stakeholders to influence the development of new regulations so they are effective, yet not unduly burdensome to the candy industry. (While the FDA is often viewed as the agency with the greatest oversight of the food industry, other agencies also play a role in food safety and include the US Department of Agriculture, the Environmental Protection Agency, the Department of Labor, and the Consumer Product Safety Commission.)
3. Track international regulations and work cooperatively with international confectionery organizations on regulatory issues of mutual interest or concern. As feasible, influence the development of other nation's regulations that may otherwise adversely impact exports of US confectionery. Work closely with members to monitor and participate in the Codex Alimentarius processes on behalf of the US confectionery industry.
4. Monitor and influence public policy and food safety issues as they relate to candy. Sometimes this means engaging with state regulatory agencies as well as federal agencies and other special interest groups.
5. Work with members to fund and oversee research related to candy.
6. Educate members and the broader confectionery industry on new regulatory requirements to foster compliance with all federal regulations that affect candy.
7. Travel as necessary (at least a few days a month) to confectionery industry technical meetings, association committee meetings, member facilities, and food industry conferences.

The job description might sound formidable, and while my job is always challenging and interesting, it is also exciting and fun. Let me share a typical day with you. In the morning I might meet with a member and tour their factory to gain a better understanding of chocolate processing and HACCP controls. I'll use this information in the future when I assess the impact of certain food safety regulations on candy production. (Of course, I accept the box of samples the generous member has offered to me on my way out the door!) Before lunch, I attend a confectionery industry meeting and update attendees on current regulatory developments and how they affect candy manufacturing. For example, FDA has recently declared that coconuts are considered a tree nut under the Food Allergen Labeling and Protection Act. This declaration impacts how candy manufacturers process and label candy with coconut or ingredients derived from coconut.

Back at my office in the late afternoon I get hungry so I sample a snack-size candy bar and a couple of sour gummy bears—both new products that members have shipped to our offices. Finally, I might round out a typical day by participating in a conference call covering a food industry–wide topic of concern, such as imported food safety and security. My objective on the call will be to work with industry association and company colleagues to share the most current intelligence on legislative developments and then determine next steps for the candy industry.

At the end of the day, I can look back and feel good that I helped members adjust to a regulation that will impact their business by providing an overview of the regulation, answering their specific questions, and offering strategies for compliance. I also learned more myself about the realities of candy manufacturing. Fortunately, I am also better prepared today than I was yesterday to put forth recommendations to the candy industry about important food safety legislation. For me, that is a rewarding way to spend the day.

What Food Science Skills Are Most Important to Success in an Association?

Assuring a safe food supply is a principle goal of all food companies and regulators. As a liaison between the regulatory agencies and candy manufacturers, this also becomes a major focus of my job. Therefore, familiarity and in-depth knowledge of food microbiology, food chemistry, engineering basics, especially as they relate to principles of food preservation and cleaning/sanitation, statistics, and nutrition are valuable skills that I use daily.

While one should strive to graduate with the highest honors in food science, applying that knowledge to real-world situations requires yet more talents. IFT lists the following skills as core competencies for success in a food science career.

- Communication skills (i.e., oral and written communication, including writing technical reports, letters and memos; communicating technical information to a nontechnical audience; and making formal and informal presentations, etc.)
- Critical thinking and problem solving
- Professionalism skills
- Interaction skills, especially leadership, interpersonal and networking,
- Information acquisition skills (i.e., written and electronic searches, databases, internet, etc.)
- Organizational skills (i.e., management of projects and consultants, facilitating groups especially long-distance teams, and conference calls)

In fact, these skills are crucial to a successful career as they allow a food scientist to put his or her food science knowledge to work in a business environment. These skills are also essential for a career in regulatory affairs as a food scientist in an association.

For a food science student entering a career in a food industry association or other special interest organization, I would recommend a strong understanding of the core food science disciplines. To supplement that education, I would suggest additional coursework, research, or summer job experience in food safety, public health, nutrition, or food law. I would also urge students to take an internship with a food company, perhaps in the area of quality control, to gain an understanding of how companies operate and how they manage food safety or regulatory issues internally. An internship with a trade association or other group located in Washington, D.C. would strengthen the resume of any candidate seeking a position in this region.

Job opportunities in associations or other special interest groups may be posted in several locations. IFT is a good place to start. The IFT Job Center is used by many of the larger associations to post job announcements and interview candidates. Additionally, the Washington, D.C. regional section of IFT covers the city of Washington, D.C., the state of Virginia, and surrounding counties in Maryland. Involvement in the Washington, D.C. section of IFT is an excellent strategy for networking with other local food technologists and food safety or health and regulatory professionals. These individuals often have the best knowledge of local organizations and employment opportunities. Food industry recruiters are sometimes retained as well to help fill vacant positions so it would be wise to reach out to those recruiters that specialize in regulatory work.

Summary

An association career allows the food science professional to work on food safety and regulatory issues that can impact the entire industry, or sectors of it, rather than just one company. With each new issue—be it an emerging safety concern, a proposed labeling requirement, or research related to the food product, there is a unique blend of stakeholders to consider including the broader food industry, regulators, other special interest groups, the media, and of course the consumer. These stakeholders and the food manufacturers themselves may have diverging goals and needs that present many challenges. The management of all those factors together while forging ahead on behalf of the membership is what makes working for an association such a unique and rewarding (and in my case, "sweet") experience.

Chapter 25
Food Science for the Public Good

Cassandra Miller

If you are interested in food science, looking for a meaningful career path, and are motivated by the desire to make a difference, you may find that a career working for the public good can be very rewarding. Often, such opportunities address issues of social responsibility, sustainability, public health, and/or economic development. Food scientists who choose this path typically have an interest in social and public health issues, and are usually driven by the achievement of some sort of social, health, or societal gain. As food science in itself is a very broad discipline, applying this knowledge for the public good can also take a variety of paths. Whether you're interested in manufacturing, food safety, nutrition, food policy, product development, quality control, marketing and sales, or any other discipline that makes up the diverse field of food science, various opportunities exist to make a difference to society.

A Personal Experience

When I entered my freshman year of college, I didn't have a clue what food science was. In fact, I didn't even know that such a major existed. I did, however, have a strong interest in health and nutrition and in helping the less fortunate. While in high school, I participated in a 30-Hour World Famine, an international youth movement to fight hunger. Having grown up with a strong interest and fascination in the sciences (I took just about every science course that my small high school offered, including advanced courses in biology, chemistry, and physics), I was determined to use my scientific

C. Miller
SUSTAIN, Washington, DC, USA

R.W. Hartel, C.P. Klawitter (eds.), *Careers in Food Science*,
DOI: 10.1007/978-0-387-77391-9_25, © Springer Science+Business Media, LLC 2008

knowledge for the advancement of public health. It wasn't surprising that a couple years later, food science became my major.

While a student at the University of Wisconsin-Madison, in addition to enrolling in course work required of any food science major, I also took courses in anthropology and nutrition. Upon entering my junior year of college, I became very interested in international opportunities and development work and started to seek internships that would give me the experience I desired. Through IAESTE (International Association for the Exchange of Students for Technical Experience), I obtained a 10-week summer internship at UNICAMP, a large research university in Brazil, where I worked on a research project studying the effects of somatic cells on the manufacturing of *queijo Prato*, a Brazilian cheese. While I learned a great deal about the process of cheese making and analytical testing methods, I was also able to gain firsthand experience living and working in a foreign country. Upon returning to Wisconsin at the completion of my internship, I immediately decided to seek additional international experience—this time in a study abroad program at the University of Pretoria in South Africa. I specifically chose this destination because I had a strong interest in development work, and this was one of the few universities that offered a program in food science.

Upon returning to the University of Wisconsin at the completion of my study abroad program, I was ready to graduate and began to seek employment. After interviewing at several food companies, I eventually came across a flyer advertising an internship position at SUSTAIN (Sharing Science and Technology to Aid in the Improvement of Nutrition), a small nonprofit organization focused on improving nutrition in developing countries through food science and technology applications. I immediately felt that this was the position I wanted to pursue, and after several phone interviews, was on my way to Washington, D.C. to begin the internship.

Most of SUSTAIN's programs involve collaborative efforts among industry, the scientific research community and governments throughout the world to improve nutrition for vulnerable populations. The office in Washington, D.C. acts as a hub for development, coordination, and administration of these various research projects. As an intern at a small nonprofit organization, my duties included administrative work in addition to project work. At the completion of the 10-week internship, I was hired as a full-time staff member.

Having worked at SUSTAIN for over a year now, I've had a chance to become very familiar with our programs, most of which are focused on enhancing the nutritive quality of food staples through micronutrient fortification. One of the projects that I've become very involved in has been our work on developing the technology to fortify corn tortillas in Mexico. Working together with partners at research universities and public health institutions in the United States and Mexico, we were able to develop a successful fortification technology that is now being used by tortilla mills in Mexico to add vitamins and minerals to freshly made tortillas.

My experience at SUSTAIN has provided me with valuable insight into operation and management of a nonprofit organization. My job responsibilities include project administration, communications, and team coordination in support of SUSTAIN's programs. I currently work with other team members to write proposals and grants and develop and manage funded projects.

SUSTAIN has helped me grow and advance my knowledge in many ways, but not in ways that have always been related to food science. Challenges encountered while working on the tortilla fortification project exemplify this. In Mexico, the tortilla industry is divided into two sectors: manufacturers that produce tortillas directly from corn flour and those that produce tortillas from fresh corn masa (dough). Initially, corn flour producers objected to mandatory fortification of their product unless similar regulations were issued for their competitors in the corn masa industry. However, no feasible fortification technology existed for the corn masa industry, which is highly fragmented and consists of many small mills scattered throughout Mexico. Therefore, SUSTAIN undertook a project to develop fortification technology for the corn masa sector. Challenges were also encountered in the development of the micronutrient premix. Not only was it important to choose an iron fortificant that was low cost, but the iron source should also provide the desired nutritional benefits (good bioavailability) without having adverse effects on the color, taste, or texture of the final product. Another challenge was to get Mexican consumers, who were not comfortable with purchasing the traditional corn tortilla that had been modified through chemical or physical methods, to accept the product. How do you confront a rooted culture about changing a traditional food staple that has been around for hundreds of years? Our solution was to design a nutrition promotion campaign to educate consumers about the benefits of fortification.

One of the biggest lessons learned from these projects is that applications of food science and technology can be extremely complex in the real world, which is often colored by changing political, economic, and cultural realities. Therefore, keeping abreast of current events, including political, economical, and cultural issues is very important.

While my interest in international nutrition and public health led me down this specific path, it's important to keep in mind that meaningful career opportunities in food science can encompass a wide range of activities and can be found in a wide variety of settings. Opportunities exist in both the public and private sectors, at research institutions and at nonprofit and international organizations. Listed as follows are a few examples of activities and organizations that fall under this domain.

Opportunities

Ensuring a Safe Food Supply

The main goal of food safety, which plays an important role in public health, is to protect the food supply from microbial, physical, and chemical contaminants during handling, storage, and preparation of food. Because improved food safety typically results in enhanced consumer well-being, longevity, and enhanced economic productivity, working for an organization that seeks to improve the safety and wholesomeness of foods can be very rewarding.

Most of the career opportunities available in the area of food safety regulation are found in the public sector and include job functions such as the inspection of foods and food establishments, the analysis of foods for contamination of pathogenic bacteria, the establishment of food safety regulations and food safety enforcement policies, the development and implementation of programs related to investigation and prevention of foodborne disease outbreaks and the development of processing technologies to assure the safety of foods.

In the United States, the two primary agencies responsible for regulating food safety are the United States Department of Agriculture (USDA) and the Food and Drug Administration (FDA). The Agricultural Research Service (ARS), the main research agency of USDA, seeks to promote the quality and safety of foods and other agricultural products. With over 100 locations in the United States, ARS hires food scientists to conduct research on microbial pathogens, chemical residues, mycotoxins, and toxic plants to help prevent contamination and to allow the food industry to make better decisions regarding safety standards and control strategies. The National Center for Food Safety and Technology (NCFST) is another organization that conducts research in the area of food safety.

Not-for Profits, such as the Center for Science in the Public Interest (CSPI) conduct advocacy activities in support of a safe food supply. CSPI's mission is to "conduct innovative research and advocacy programs in health and nutrition, and to provide consumers with current, useful information about their health and well-being." The CSPI's Food Safety Program provides current research on food safety issues to the public, policymakers, and regulators. The program encourages USDA and FDA to strengthen food safety programs and lobbies Congress to strengthen food safety laws. For those interested in consumer advocacy and public policy development related to food safety and nutrition, pursuing a job at a group such as CSPI or The International Food Policy Research Institute (IFPRI) might be a good fit. Having a solid understanding of the disciplines related to food science, public health, and nutrition would be very valuable. Depending on the position, however, employment opportunities at CSPI, IFPRI, and other policy-related organizations may require experience in state/federal legislature, policy research, or policy analysis.

Food Aid Products for Humanitarian Relief Programs

Developing new food products that can be easily stored, distributed, and utilized in emergency situations as food aid can be critical to ensure food security of populations that lack access to food due to natural disasters or poor economic situations.

In the United States' *Food for Peace Program (P.L. 480)*, food aid commodities such as rice, wheat flour, cornmeal, kidney beans, lentils, and vegetable oil are purchased by USDA from commercial food manufacturers and then distributed overseas by private voluntary organizations (e.g., World Vision, CARE, Catholic Relief Services) and the World Food Program (WFP). These food aid commodities are typically manufactured in the United States by commercial food manufacturers such as Cargill, Bunge, Didion, or ADM. Within these food manufacturing companies, food scientists are hired for product development, production management, and quality assurance positions. Since food aid is typically a small part of these companies' overall business, however, opportunities to work on humanitarian projects may be limited. Consequently, much of the basic food science research in the area of international and humanitarian development takes place in international organizations, nonprofit organizations, academia, and/or governmental settings.

SUSTAIN has done a lot of work in the food aid arena, working with food scientists from various sectors (including industry, government, and academia) to enhance the nutrient delivery of food aid commodities through improved quality control and updated formulations. The SUSTAIN team includes in-house staff, consultants, volunteers, and partners in developing countries with expertise in project management, food science, and nutrition. SUSTAIN teams have been involved in assessing the micronutrient quality and enhancing the nutrient delivery of food aid commodities through improved quality control and development and revision of commodity specifications. Research being conducted by SUSTAIN and other organizations has also indicated that adjustment of nutrient formulations may be necessary to better meet the nutritional needs of target groups. Currently, many of the food aid products are distributed to beneficiaries as one-size-fits-all products; however, certain vulnerable groups, such as infants and young children, pregnant and lactating women, and people living with HIV/AIDS have very different nutrient needs. Reformulation opportunities being explored include adjusting the levels of certain vitamins and minerals in the micronutrient premix and adjusting ingredient formulation to improve energy density, alter viscosity, improve protein quality, and extend product shelf life.

The World Food Program (WFP) is the largest humanitarian organization in the world, distributing about four million tons of food on an annual basis. Optimizing the nutrition and quality of food aid is a major concern for the activity of WFP. In addition to dealing with nutrition, food quality, and food safety, WFP plays an important role in the purchase, storage, transportation, processing, and distribution of food. WFP hires people experienced in food processing and quality control to prevent the risk of foodborne illness, reduce losses related to product recall, develop local food production, and reduce costs of food analyses. All of these activities related to food quality control play a crucial role in improved health and potential for a country's economic development. Typical job functions of a quality control specialist at WFP might include identification of losses of commodities in the food pipeline, development of training modules on loss prevention for logistics officers, collation of data on the cause of commodity losses, and formulation of loss control measures.

Providing Technical Assistance to Local Food Manufacturers in Developing Countries

Many food manufacturers in developing countries lack technically trained staff, equipment, and knowledge about new food processing technologies. Qualified laboratories specializing in food analysis are usually scarce and governments often lack the financial and technical resources necessary to ensure a safe food supply. Providing technical assistance and training to local food industries in these areas can help improve food safety and strengthen the capability of these businesses to supply quality products on a sustainable basis.

The Food and Agriculture Organization (FAO) and the World Health Organization (WHO) are involved in technical assistance projects aimed at strengthening food control systems in developing countries. Through information dissemination and knowledge brought directly to the field, FAO and WHO provide assistance to developing countries on preparation of food legislation, regulations, and standards; training and implementation of food inspection programs; development and improvement of food analysis capabilities; training in management of food control systems; and promotion of technologies designed to prevent foodborne diseases.

In the context of food aid, the purchase of local foods (foods that are produced, processed, distributed, and consumed within a given region) not only benefits the local economy but also leads to decreased food storage and transportation needs and results in a specific formula adapted to beneficiaries needs and tastes. The World Food Program purchases the majority of its food in developing countries, often relying on surveillance of the manufacturing plants and/or analysis of the finished product to ensure food product quality. WFP plays an important role in developing local food manufacturing capacity by supporting government food legislation, promoting best practices at factories, and providing technical assistance to food processors on food fortification and quality control.

One of the greatest contributions that could be made by a food scientist is the establishment of new food processing facilities in developing countries, which contributes to economic wealth, food security, and job creation. Local citizens will have the opportunity to gain firsthand experience in food processing technologies and quality control systems, generating a body of knowledge and experience, which can be passed on to future generations.

What You Can Do to Prepare for a Career in Food Science Serving the Public Good

Most organizations seeking food scientists employed for the public good also seek candidates that are skilled in areas beyond that of their primary food science background. Additional experience in project management, computer applications, statistical analysis, contract administration, grant and proposal writing, finance, and recruitment/networking are all valuable skills. However, because the field is broad and the opportunities vary greatly, the specific experience desired will depend largely upon the type of organization and position sought.

In terms of course work, it is essential to have a basic understanding of core areas that make up the discipline of food science, including food processing, food microbiology, food engineering, and knowledge of the major food categories (fruits and vegetables, cereal grains, dairy, meats, and seafood). The ability to effectively communicate food science issues to people in government, the health sector, and the manufacturing sector in a nontechnical language is essential, as many people working in these organizations lack scientific backgrounds. Therefore, courses in writing, specifically scientific writing, would be useful. Those interested in food policy may find courses in political science, policy analysis, or public administration helpful and may even wish to pursue a dual or specialized degree. For example, at the University of Massachusetts Amherst, the Center for Public Policy and Administration and the Department of Food Science have combined to create a program in Food Science Policy. For those interested in international and humanitarian opportunities, courses in nutrition (especially international nutrition) and international development would be particularly valuable. In addition, proficiency in a second language is sometimes required for those seeking positions at international organizations such as WFP or WHO. Another important attribute is flexibility and sensitivity to cultural issues. Courses in economics, business, and marketing may also be useful. In general, it is a good rule of thumb to have depth and expertise within one specialty, with hands-on experience, and/or knowledge in another specialty to broaden your scope.

Gaining worthwhile experience is a highly valued asset. Some opportunities may be available right in the food science department at your university, as faculties are often encouraged to hire undergraduate students for assistance on research projects. Professors can also be a great source of knowledge about available opportunities, and may be able to put you in contact with an organization that specializes in the type of work you are seeking. Even if you are unable to obtain an internship with an organization that specializes in humanitarian work, securing an internship at a food manufacturing company is highly desired, as international and humanitarian organizations seeking food science specialists are typically interested in those with industry experience. Consequently, the best approach might be to get some general experience in the food industry for a few years before trying to get into humanitarian research. That way you can draw on both your academic as well as your industry experience to help solve problems with food safety and world hunger.

For those interested in international opportunities, obtaining international experience is highly recommended. Several programs such as AIESEC and IAESTE (International Association for the Exchange of Students for Technical Experience) offer international internship exchanges. Participating in a study abroad program is another way to gain international experience. If possible, choose a university that offers food science and a country that has a need for the type of work you desire so that you can familiarize yourself with the local cultural and political environment.

Another way to gain valuable real-life experience is through Peace Corps. As a Peace Corps volunteer, project activities could include anything from teaching chemistry to high school students to helping farmers improve local diets and increase income through farming techniques. The Peace Corps can be a great stepping stone to future opportunities at international organizations. Resourcefulness, creativity, international experience, and leadership are some of the strengths that Peace Corps volunteers develop during their service. The challenges of limited resources and poor communications are commonly faced by Peace Corps volunteers, and are a great preparation for the realities of conducting food science work in the developing world.

Several universities offer the Master's International Program, which combines a master's degree with overseas Peace Corps service. Even for those not interested in an international career, Peace Corps service can be a great opportunity to do some volunteer work before joining the food industry.

Summary

Working in the field of food science and technology to serve the greater public good can be a challenging yet rewarding career. One of the greatest rewards for working at an organization that serves the public good is the fulfillment that comes in knowing that you are helping to make the world a better place, whether through economic development, sustainability, improved public health, or any other area that makes a difference.

Chapter 26
Careers that Combine Culinary and Food Science

Michelle Tittl

Imagine yourself perusing the aisles of your local grocery store. You head down the frozen food section and, being a cost-conscious shopper with little to no time to cook, you choose a seemingly delectable heat-and-serve meal of grilled chicken medallions and sautéed spinach doused in a mushroom sauce. Taking a closer look at the bag, you ask yourself, is this a delicious food concoction of culinary art or of food technology?

A Taste For More

Traditionally, food scientists were once the main creators in the development of food products for grocery store items. However, today's consumer knowledge of various types of exotic food and ethnic cooking techniques has greatly expanded over the years, including their taste for more sophisticated, gourmet meals. One cause may be due to an increase in the cooking resources available on television, books, the Internet, and the awareness of celebrity chefs through the media (Doyle, 2006). Even, independent chefs, such as Grant Achatz of Alinea Restaurant and Homaro Cantu of Moto Restaurant both in Chicago, Illinois have earned a reputation for embracing a scientific approach when crafting new menu items that include foams, vapors, and even edible paper for guest's amusement that leave them wanting more (Molecular Gastronomy Resources, 2007). The food media has also picked up on terms such as "molecular gastronomy" and "techno-chef" that have been used to describe this science-based approach of food

M. Tittl
Target Corporation, Minneapolis, MN, USA

R.W. Hartel, C.P. Klawitter (eds.), *Careers in Food Science*,
DOI: 10.1007/978-0-387-77391-9_26,

innovation in experimental-style kitchens that incorporate new equipment and techniques within the restaurant industry. As a result, the food industry has taken a fresh approach to designing food products by incorporating more culinary perspective into product development in order to be on the cutting edge in today's marketplace (Doyle, 2006). Currently, food manufacturers are well aware of this need and have begun hiring research chefs with a background of both food science and culinary arts to create restaurant quality recipes that can be commercialized. In addition, more certified chefs are now crossing over to become involved in food manufacturer's research and development (R&D) process to provide flavorful, innovative ideas for savvy consumers (Cooper, 2006). The blending of food science and culinary arts results in an understanding of not only manufacturing processes, ingredient functionality, and food safety requirements, but also a culinary edge that can improve the flavor and overall appearance of product concepts (Doyle, 2006).

The Rise of the Research Chef

Once the number of research chefs within the food industry began to grow, a group of food professionals were determined to create an organization that could provide a support system for food specialists with a common interest and address the many challenges facing their particular career path. Formed in 1996, the Research Chefs Association (RCA) was founded by Research Chef Winston Riley and has now grown to over 2,000 members becoming an excellent source for culinary and technical education and information for its members within the food industry. The RCA membership includes chefs, food scientists, salesmen, and other food professionals within food manufacturing facilities, chain restaurants and hotels, supply companies, consulting services, and academia (Research Chefs Association, 2007).

Emphasizing the need for a culinary point of view in the food business, Riley, then president, wrote in a 1997 RCA newsletter, "Our future is bright, considering the great importance of the culinary perspective in the development of products and services, as we approach the new millennium" (Cousiminer, 1999). From this, Winston Riley later coined the term "Culinology" to describe and formulize the fusion of the food sciences with culinary art. This term, now trademarked, recognizes the birth of a new expertise within the food industry that produce restaurant quality foods on a manufacturing level that look and taste like food served in a restaurant. The continual interest in Culinology has created new educational opportunities through various universities, certificate programs for professionals, and specialized courses available for individual growth and development (Research Chefs Association, 2007).

A New Flavor in Education

Until recently, the fusion between culinary arts and food technology was not available for students interested in food-related careers. But today, more educational institutions and universities are aware of the growing interest in combining this knowledge throughout the food industry and therefore, have provided more opportunities for students who are interested in food and have both creative and analytical talents. In an interview, Danny Bruns, corporate chef for Beloit, Wisconsin-based Kerry Americas, and a founding member of the RCA stated, "Education is really the big push, getting students to understand both sides of the business. Culinology, combining the two paths, is really the next wave in food creation" (Cooper, 2006). Currently, colleges and universities have begun to offer classes that teach culinary fundamentals for food scientists as well as educate chefs on food science concepts. Some university food science programs partner with local community colleges' 2 year culinary degrees. Other major U.S. universities offer traditional 4 year food science programs that provide some culinary background (Hahne, 2001).

Today, the RCA has eight approved "Culinology" programs within various universities and colleges across the United States and continues to expand additional institutions. These approved programs offer the majority of classes in culinary arts and food science, as well as supporting curriculum in chemistry, business management, nutrition, processing technology, and government regulations to provide students with a well-rounded education in the "Culinology" program (2007). The University of Nebraska-Lincoln (UNL) was the first international-approved RCA degree program in "Culinology." Since then, other approved degree programs among universities include: California Polytechnic Pomona/Orange Coast College, California State University Fresno, Clemson University, Dominican University/ Kendall College, Southwest Minnesota State University, University of Cincinnati/ Cincinnati State Technical and Community College, and University of Massachusetts-Amherst (Research Chefs Association, 2007).

Additional Culinary Certification

Another option that students have had in the past is to continue their education in the culinary field after obtaining their BS in food science by attending either a 2 year associates degree or a 1 year advanced culinary certificate at an accredited culinary school. Top-ranked culinary schools including The Culinary Institute of America (CIA) and Johnson and Wales (J&W) as well as local community colleges provide outstanding 2 year associate programs that incorporate a 6-month externship with hands-on experience in either hotels/restaurants or food manufacturing facilities.

The CIA also provides an Advanced Culinary Arts Program Certification (ACAP) at their Saint Helena, California location. This option is meant for individuals who already have acquired a food-related degree and have a passion for the culinary arts! The ACAP program provides the same courses as an Associates program without the 6 month externship and is condensed into an intense 8 month curriculum that is fast paced and vigorous in study. Both the certificate and degree programs include courses that cover basic knife skills and cooking techniques, production cookery, various world cuisines, flavor dynamics with ingredient interaction, and techniques of healthy cooking (The Culinary Institute of America, 2007).

Culinary school in general is extremely costly, with an estimated tuition somewhere around $30,000 per year. If interested in any of these programs, students who already have a BS in food science might opt for the concentrated Accelerated Culinary Arts certificate, which is only 1 year long as opposed to the 2 year Associates program to save on costs. However, the 2-year programs devote more time to each of the courses and provide more experience including the externship within the food industry. Keep in mind, it is extremely advantageous to gain experience in a restaurant or hotel service before attending culinary school. This opportunity will provide a better perspective of how food is prepared in an industrial setting as well as acquire some skills that will put you ahead of other culinary students with no experience (Doyle, 2006).

With all of the various career paths available, students have the opportunity to gain a better understanding of how both food science and the culinary arts are related and interact simultaneously. This results in a more valuable candidate within the food industry specifically in careers such as R&D and sensory science. Theoretically, when potential employers interview job candidates with continued education, they look at the applicant more favorably (Luff, 2002). Certainly, this depends on the degree of certification and how applicable it is to the job opportunity. With the ongoing trend of the culinary/science combination, more employers will be seeking individuals with this collective knowledge. As an example, John Kennedy II, a former culinology degree student, has begun working as a Culinologist for Well's Dairy, Inc., Le Mars, IA upon graduation, and states in an interview in a "Science in the Kitchen" article that, "A culinologist is a valuable position that strategically places you between a chef and a food scientist in a manufacturing setting" (Doyle, 2006).

Choosing the Right Recipe for Your Career

Once out of school with a background in both food science and culinary arts, students may find a wide variety of opportunities that may not have been available with just one food-based focus. These opportunities may include a R&D chef, a culinary scientist, or even a corporate executive chef (Research Chefs Association, 2007). According to a study from the Purdue University of West Lafayette, Indiana, there were found to be two different types of opportunities that practice the unique blend of art and science throughout the food industry that possesses a certain set of attributes. One type is considered a research-based position that has strong culinary knowledge and food ingredient understanding. This individual is primarily involved in commercial development of new products and processes. These opportunities include product developers within R&D departments of food companies who develop grocery store items that may be shelf-stable, retorted meals, frozen and refrigerated meals, dry mixes, and frozen microwavable finger foods. A strong culinary background is extremely helpful in developing creative new food products within these categories, as well as understanding food processing equipment and methodologies. The second type of position is management focused; it is someone who works alongside customers, conducts sales presentations, and is involved in the strategic planning process of food production. These opportunities may include a culinary scientist or corporate chef who works alongside the sales team as a consultant to provide culinary insight to other food companies in order to assist in initial product ideation and provide feasible gold standard recipes for manufacturers (Culinology program overview, 2007).

A wide variety of food companies are beginning to show more interest in hiring individuals with a culinology background. Depending on previous experience, food companies, small to large, are continuing to find recent graduates with the art and science combination as an advantageous addition to their staff in order to generate new ideas and creative culinary concepts. Even retail companies such as Target Corporation, Whole Foods, and Tesco have been interested in hiring individuals with a culinary/science combination to aid their vendors in the development of new private label food items. For example, Target Corporation has recently hired graduates from the culinology program at the University of Nebraska and the Culinary Institute of America Advanced Culinary Arts Program, as well as establishing an internship program through prestigious culinary schools to work alongside other food scientists and create premium gourmet products for their Market Pantry and Archer Farms private label brand.

After recently being hired at Target Corporation as an Associate Food Scientist, I have had the opportunity to work alongside other product development scientists as well as with a variety of food companies to provide successful, great tasting products within our owned brands. Since our branded products cover a variety of categories including snacks, frozen, confectionary, and much more, I find myself working on an assortment of food projects where I can assist in the development of a granola bar one day and work on cream-based sauces the next. By having a background in both food science and the culinary arts, I have the ability to not only understand the technical aspect of a product, but I am able to provide insight in unique flavor profiles, innovative cooking techniques, and global ingredients that could enhance the overall product for our guests to enjoy! Overall, this combination of both art and science creates a valuable position for food manufacturers and retailers to utilize your skills equally, effectively, and successfully. Through my experience, I recommend this career path for anyone who has a true passion for all aspects of food!

Ingredients for Success

When seeking potential candidates, companies are looking for specific competencies that will best fit a position who can offer a competitive edge through the knowledge of art and science. A study for the RCA found that their members all contain a certain set of core competencies that best describe what it takes to fulfill a culinology job description. Members were found to exhibit a set of knowledge competencies, skill and ability competencies, and behavior competencies. Some of the most significant competencies are the knowledge of various flavors and their interactions when blended together, innovative recipe development, and the ability to create "Gold Standard Recipes," as well as a full comprehension of ingredient functionality. An understanding of foodservice operations and food production systems, which include appropriate food preparation, essential equipment, and proper sanitation requirements, are also important in order to provide well-rounded experiences. In addition, a culinary education, a scientific or technical education in food science or food management, and food processing knowledge through "on the job" experience (Culinology program overview, 2007) are critical for development of the skills listed above. Most importantly, food manufacturers are looking for individuals with a sense of taste and a passion for food!

Unique Credentials

Once established within the food industry, qualified professionals have the opportunity to obtain unique credentials from the RCA. These include the certification for Certified Research Chef (CRC) and/or Certified Culinary Scientist (CCS). The professional credentials are highly respected within the food industry around the world and add more value to candidates pursuing opportunities for their proven skills and experience. Both certifications are available to members and nonmembers of the RCA (Research Chefs Association, 2007).

The CRC is available to *culinary professionals* who have a specific amount of experience working in food product R&D. Applicants must meet certain eligibility requirements in order to receive this specific certification. These requirements include a bachelors or associates degree in culinary arts, a certificate in culinary arts, or 30 hours of nutrition, food safety, and culinary professional development. Depending on the type of education that has been achieved individuals are required to have at least one full year (2,000 hours) of cooking in a production or supervisory position in a commercial kitchen. All routes require at least 3 years of full time (2,000 hours per year) experience within product development and require a written CRC validation exam that is largely based on food science knowledge, in addition to culinary arts (Research Chefs Association, 2007). The RCA explains that CRCs with this accreditation are "among the most knowledgeable in their field and are leaders in the food industry because they have proven competence in both culinary arts and food product research and development" (Research Chefs Association, 2007). This certification provides a verification of experience, education, and expertise that enhances the value that an individual can bring to a particular company.

The Certified Culinary Scientist (CCS) is available to *food science professionals*. Similar to the CRC, individuals must meet certain requirements in order to be presented with this highly recognized certification. These requirements include a BS degree or higher, an Associates Degree, or 30 hours or more of college level courses in microbiology, nutrition, and chemistry. Depending on the type of education achieved, individuals must have at least 3 years (2,000 h each) of product development experience. All routes require 1 year of full-time production and/or supervisory experience in a commercial kitchen or must pass an American Culinary Federation Culinary Practical Exam. This certification also requires a written CCS validation exam that is largely based on culinary arts and a small portion on food science (Research Chefs Association, 2007). The RCA states that the CCS certification provides "a new status on experienced food scientists and technologists who have augmented their training by learning about the culinary arts and who use this knowledge in the development of superior food products" (Research Chefs Association, 2007). The CCS certification provides recognition to valuable candidates through proven work experience, education, and expertise that are also seen as great assets to food manufacturing companies.

Professional Organizations

When first beginning the educational experience as a student in either food science or culinary arts, it is crucial to be a part of professional organizations. These opportunities allow students to not only keep up with current food trends and new technological processes within the food industry but to utilize membership as a networking tool to build relationships with various companies and individuals in order to gain a better understanding of what career path is applicable to you. In addition, involvement in college-based clubs and organizations such as food science clubs, culinary clubs, or the product development club will link you to affiliated national organizations that will also help build connections for the future.

The RCA is a wonderful tool for anyone who is interested in pursuing a career that combines the culinary arts and food science. Membership dues for students are only $25 annually and include scholarship opportunities for individuals in food science, culinary arts, and culinology programs. In addition, the RCA holds an annual conference and Culinology Expo, which is a great resource to meet industry professionals and learn from individuals who have established careers practicing the culinary arts and food science. For more information on the organization and additional resources, refer to their Web site: www.culinology.com (Research Chefs Association, 2007).

The Institute of Food Technologists (IFT) is also a wonderful organization to join as a student. With over 22,000 members, this nonprofit organization has members in food science and technology, academia, and the government that can assist any student in discovering opportunities within the food industry. Student memberships are only $35 per year and include a subscription to either the *Food Technology* magazine or the *Online Journal of Food Science*. In addition, all student members must be endorsed by a faculty member. The IFT also provides a variety of local and national scholarships for student individuals studying food science or any food-related major. It also conducts the world's largest annual food convention known as the IFT Annual Meeting and Food Expo that provides students with a great opportunity to network among professionals and learn from their experiences (Institute of Food Technologists, 2007). More information and resources are at the IFT Web site: www.ift.org. Overall, as a student, it is extremely important to get involved in various organizations within the food industry in order to acquire a career path perspective and develop into a more professional candidate that will appeal to companies upon graduation.

Summary

Over time, the marriage of culinary arts and food science has proven to be beneficial and valuable in today's current food manufacturing world. Today, there are over seven accredited culinology programs across the United States as well as additional continuing education programs that can help launch your educational career in the art and science of food! In addition, continual participation in professional organizations and clubs will greatly assist you in making both industry and educational contacts in addition to providing you with a better career path perspective. As job opportunities within the food industry continue to grow, candidates with this knowledge have a competitive edge over an individual with just one focus, providing more perspective when designing new products. Hence, upon examining that frozen delectable chicken entrée while shopping at your local grocery store, you now know that this particular product is both a combination of culinary arts and food science.

References

Cooper, Carolyn. "A Fine Balance." June 2006. Research Chefs in Canada. August 2007. www.foodincanada.com.

Cousiminer, Jeffrey. "Practicing Culinology." January 1999. Food Product Design. August 2007. www.foodproductdesign.com/archive/1999/0199 cc.html

"Culinology program overview." University of Nebraska-Lincoln Nutrition and Health Sciences. October 2007. www.cehs.unl.edu/nhs/undergrad/culinology.html.

Doyle, Anneliese. "Science in the Kitchen." June 2006. Research Chefs in Canada. August 2007. www.foodincanada.com.

Hahne, Bill. "Education Benefits Chefs, Food Formulators-Guide to Culinary and Chef Techniques." December 2001. Prepared Foods. August 2007. www.findarticles.com/p/articles/mi_m3289/is_12_170/ai_80749070/pg_1.

Institute of Food Technologists. August 2007. www.ift.org.

Luff, Steven. A. "The Push to Professional-Certification Programs." October 2002. Food Product Design. August 2007. www.foodproductdesign.com/archive/2002/1002RD.html.

Molecular Gastronomy Resources. October 2007. www.alacuisine.org/alacuisine/2004/11/molecular˙gastr.html.

Research Chefs Association. August 2007. www.culinology.com.

The Culinary Institute of America. August 2007. www.ciachef.edu.

Chapter 27
Food Business Entrepreneurship

Peter Weber

Introduction

Though not a very traditional career path for food scientists, one option is to go into business for yourself by starting a food business. Food business entrepreneurship is a difficult career that entails long work hours, extensive decision making, and tasks that require knowledge beyond food science. However, there is high potential for rewards, including financial rewards, career progression, and personal flexibility.

Types of Food Businesses

Food science is a multidisciplinary field that applies the basic sciences of food systems to appease the consumer's stomach. The broad range of studies within food science and the gigantic size of the food industry allows for many types of businesses to be developed. Food scientists are not limited to starting a food processing company. Following is a list of areas of the food industry in which a food scientist could think about starting a business:

- Ingredient supply
- Raw material management
- Specialty processing
- Consulting
- Packaging
- Research and development
- Sensory analysis
- Quality control

P. Weber
Potter's Crackers, Madison, WI, USA

R.W. Hartel, C.P. Klawitter (eds.), *Careers in Food Science*,
DOI: 10.1007/978-0-387-77391-9_27, © Springer Science+Business Media, LLC 2008

- Product distribution
- Food production
- Food marketing

The type of business you decide to open should be based on an opportunity that you identify, your knowledge of an area of the food industry, and your particular interest.

Steps in Starting a Business

There are a couple of critical steps in starting a business. The first step is idea generation. The idea for a business usually comes form either a random thought (I just had a good idea!) or from analysis of a market. People who know they would like to start a business, but do not have an idea, usually do a market analysis study. The analysis can help them identify an idea for a business. The details of market analysis are beyond the scope of this book but can be found in many books on the subject of marketing. During this step you should be thinking about and developing your idea. It is recommended at this phase that you research the concept and that you discuss it with trusted friends and colleagues in your social network. This will allow you to begin to refine the idea through feedback.

The next step in developing a business is to write a business plan. This is a very crucial step that has a huge impact on the success of your business. Your business plan is the blueprint of the business that acts as a model for the business opportunity. The business plan should be used as a tool to evaluate the business opportunity before any capital is invested. If your finances do not work in the business plan, they will not work in reality. Many good books exist on the topic of writing a business plan and you can access them through your local library. Once your business plan is written, you should have trusted associates from food science, finance, and business look over the plan and discuss it with you. Have them look for gaps or holes in the plan and then edit the plan. It is usually cheaper to make a change in your business plan than it is to make a change in an existing business, so make sure you get a lot of feedback on your plan. Once your business plan is finalized, you are ready to execute it. With a comprehensive business plan in hand, execution should be smooth.

Preparing to Be an Entrepreneur

Since food science is such a multidisciplinary study, food scientists tend to have a diverse set of skills, which is helpful when opening a business.

Nonetheless, skills that are often not part of a food science curriculum are needed. These include:

- Reading accounting statements: balance sheet, cash flows, and income statement. (Managerial accounting 101)
- Basic human resources/ hiring practices (Understand laws associated with hiring so that you do not get sued)
- Standard sales and marketing practices (Business etiquette)
- Basic understanding of operations planning (number of employees, production rates, number of shifts, bottlenecks)

Many of these skills can be attained through readings and talking to professionals. If you do not have a good understanding of one of these areas, one option is to hire a consultant who specializes in the area of concern.

Owner's Organizational Role

Your organizational role is what defines your position within the organizational structure and assigns you the tasks for which you are responsible. The business owner usually takes on the role of business leader but, as a food scientist, you could also take on a more technical role within the organization and leave the business processes to business specialists. This is something that should be considered and discussed in the business plan.

Entrepreneurship Resources

There are many resources for food businesses entrepreneurs. They can be categorized into main two groups; resources for starting any kind of business and resources specifically for food businesses. The resources can come in many forms including advising, financial, networking, and more. Here are a few examples:

- Small Business Development Centers—on college campuses around the country
- Private consultants
- University extension programs
- Trade groups
- Local economic development organizations
- Business incubators/accelerators
- State government programs
- Federal government programs

- Books published on the subject—check with your local librarian
- Entrepreneurship clubs
- USDA
- FDA

An important resource for starting a business is other businesses. Much can be gained by observing similar businesses and how they operate. These are called role-model businesses and you should try to create a group of role models for your businesses. These businesses do not have to be your competitors but can be businesses that share a similar business process. It is sometimes easier to get information from noncompetitive businesses; if you are not a competitor you can commonly contact them, sit down, and talk to them.

Potter's Fine Foods

Potter's Fine Foods is an organic foods company located in Madison Wisconsin. It is a food processor that makes organic food products for sale in specialty food stores. It produces products for sale under one of its own brands, Potter's Crackers, and also contract produces for other companies.

I grew up in a family that owned a food business and when I came to the university I knew that I wanted to open up a business of my own. I started the company during the last year of my undergraduate program, while pursuing a degree in food science with a focus in business at the local university. My degree in food science with the addition of the business focus properly prepared me for operating a food company.

When I started Potter's Fine Foods, I had to go through the same process that was described above. I started out with idea generation. I did not have to do a formal market analysis because I had a few ideas for product. I made the products and brought them to class and handed them out for everybody to try. There I got a lot of feedback on product formulation and also the business idea. Once the idea had been in my head for about a month, I was ready to start writing a business plan. Since this was not something I had done before, I had to research how to write a business plan. The book that I used and would recommend is *Writing a Convincing Business Plan—second edition.* (DeThomas and Grensing-Pophal, 2001). My business plan ended up being around 25 pages. Once the plan was written, reviewed, and edited, I executed the plan. This was the easy part because it took very little thought—all the thought had been done in developing the plan. The time between when I thought of the idea and when the first product started rolling off the processing line was about 4 months.

When I started Potter's Fine Foods, I used a few of the resources listed above. I used a number of books including the book listed above to help write a business plan. I started the business using space rented from a non-profit business development group. This allow for reduced rent and shared resources, like forklifts and freight elevators, in exchange for providing good jobs for the neighborhood. I consulted the local Small Business Development Center to find out about the laws and regulation that I need to follow.

I knew that I want to be an entrepreneur when I came to the food science program and it has, with addition of some business courses, adequately prepared me for starting a food business. A deep understanding of food, which the food science program has given me, is crucial in continuing to produce safe product of the optimum quality. Due to the biological nature of foods, there is always variation within the products and their raw ingredients. This leads to problems that must be dealt with on a regular basis, like product reformulation quality testing. Without a degree in food science this would not be possible for my company. I was also able to take about five business courses as an undergraduate and they have been very valuable for me as an entrepreneur. The courses that I took were all basic first level courses like marketing, accounting, operations, human resources, and entrepreneurship. Without them, I do not think I could run a successful business.

Summary

Starting a business can be a great thing but it is also hard work that requires the entrepreneur to wear many hats. With such a large industry, you have many choices on types of business to start. To find your niche within the food industry, you can do a formal market analysis or can come up with a good thought. The process of starting a business should go smoothly as long as a good business plan is written, reviewed, and edited. Take advantages of all of the resources offered for entrepreneurs.

Reference

DeThomas, A.R. and Grensing-Pophal, L. 2001.*Writing a Convincing Business Plan.* Barron's Educational Services, Newyork.

Part VI
A Successful Academic Career

Chapter 28
How to Land the Academic Job

Silvana Martini

Introduction

Looking for a job is not an easy task. It is time consuming and, most of the time, frustrating. Academic positions are no exception to the rule. Looking for an academic position is as demanding and exciting as any other job search. To minimize frustrations and surprises, the job search should be approached with the right attitude and preparation. The first step toward finding a job in academia is to understand the implications and responsibilities of an academic job. Realizing an academic job is the future that an individual wants to follow is a difficult decision. For most people it is not an easy choice to make. A significant amount of time and effort must be devoted to the job search process; therefore, narrowing the possible choices by selecting specific areas of interest is a good strategy to pursue.

As in any job situation, networking is a fundamental tool to start building a successful professional career. Creating contacts and learning how to interact with people is an important quality, which needs to be acquired and practiced at the very early stages of a career. Needless to say, professional contacts always play a fundamental role in the job search experience. However, networking by making and maintaining contacts is not enough. Hard work and productivity are the best indicators of a successful professional. Experiences gained during the first years in academia as a graduate student are important in establishing the necessary background that will be drawn on for the rest of a professional career. Finding collaborators in a specific area of interest is a perfect combination between creating a good network and working hard. Collaborations with colleagues will increase personal confidence

S. Martini
Department of Nutrition and Food Sciences, Uath State University, Logan 84322–8700, UT, USA

R.W. Hartel, C.P. Klawitter (eds.), *Careers in Food Science*,
DOI: 10.1007/978-0-387-77391-9_28,

and professional credibility among the scientific and private communities. Besides collaborations, postdoctoral positions are also a good option. Postdoctoral experiences outside the institution where the PhD was obtained are unforgettable and contribute to growth as an independent professional. Taking the opportunity to experience new techniques and different research styles or philosophies helps to broaden the spectrum of research possibilities and strategies that can be used to develop a successful professional career. Academic positions typically include a high percentage of teaching responsibilities; therefore, early development of teaching skills is important and generally valued by the academic search committee. In summary, obtaining good research and teaching qualifications are what indicate an exceptional candidate.

Some general tips for the search process are:

- *Be yourself:* When speaking to people, be yourself. Trying to be somebody you're not, only results in wasting your time and that of your potential employer.
- *Be specific and know what you want:* People like to hire enthusiastic individuals with clear goals, who won't waste time in getting started since they have a clear big picture of the trajectory that they want to give to their professional career. A good approach is to face the job search with clear teaching and research strategies and philosophies; this will not only help to build a successful application material, but also to the development as an academic professional.
- *Adopt a positive attitude:* Interviews should be faced with a positive attitude. Face the interview with confidence and with the objective to do a good job and give your best answers to the search committee. If you follow these rules, even if you don't get the job, you will cause a good impression in your colleagues and it is very likely that a new network contact will arise from the interview.

No matter how discouraging the process might be, always remember what the objective is: *find the right job.*

Making Up Your Mind

Before jumping into the job search process, the type of academic position desired must be established. To make the right decision, it is essential to understand the different types of academic appointments and the responsibilities entailed to each.

A career in academics might involve several different areas such as teaching, research, and extension. Depending on the academic institution's needs and on personal preferences, appointments may consist of only one (i.e., teaching, research, or extension), any combination of two (i.e., teaching and research, teaching and extension, extension and research), or all three of these fields (i.e., teaching, research, and extension). A brief description of the responsibilities of each academic area is presented as follows:

Teaching usually plays an important role in academic positions. Appointments can vary from 0%–100% teaching depending on the institution's need. While teaching responsibilities include both undergraduate and graduate courses, the supervision of graduate students as either a major professor or a committee member is also considered as a teaching task. If the position includes a teaching effort, then incorporating a *teaching portfolio* with a detailed *teaching statement* and *philosophy* in the application material is a good idea. Information about building teaching portfolios can be found at Bullok and Hawk (2001); Constantino and DeLorenzo (2006); Hutchings (1998); Mack-Kirschner (2005); and Seldin (2004).

Research is another area usually related to academia. Both basic and applied research is of interest to academic institutions. The responsibilities of an academic researcher are to establish an externally funded program, either federal or private, and to gain unique expertise in a specific area of research that will benefit the local community and society in general. If a specific appointment includes research efforts, then the job interview is going to be focused on the ability of the candidate to build an independent research program.

Extension professionals represent a link between research performed at the university and the community. The role of "the cooperative extension system" is to extend university resources to people and improve the quality of life for individuals, families, and communities. Extension is usually funded by federal, state, and local governments. The extension system is composed not only by state universities but also by a network of county offices. Over 3,000 counties across the country participate in extension programs; however, university faculty and staff may build local programs to deliver specific educational tools according to the community needs. This integration of teaching, research, and public service enables the cooperative extension system to respond to critical and emerging issues with research-based, unbiased information (http://www.extension.usu.edu/htm/about).

In addition to teaching, research, and extension, when joining an academic institution, every single employee becomes part of a "big family" and as such, service activities are usually included under professional responsibilities. Service activities may include being a part of search committees, tenure and promotion committees, and participating in general activities offered by the department and/or the university.

Making up your mind about the direction of your professional life is the first thing to be done before even starting the job search. To make the academic job search more efficient, I suggest taking the time to find out what your strengths and interests are. Additional information about academic job expectations can be found in Bataille and Brown (2006) and Baez and Centra (1995).

Finding the Right Advertisement

Looking through the different job search resources is not as easy as it seems. Using some common sense and keeping in mind specific areas of interest makes the process significantly easier. To avoid feeling overwhelmed and frustrated, a good strategy to follow is to focus on a specific job description. Narrowing down and selecting only the jobs of greatest interest saves time and effort for both the applicant and future employer. Resources that usually post academic job advertisements are scientific magazines, newspapers, and Web pages. In addition, professional contacts might also provide a good source of information regarding the current academic job market.

Preparing the Application Material

As mentioned previously, the first step to a successful job application is to determine a specific area of interest. After this fundamental decision, the next important step is to look for advertisements that describe those interests. Last, but not least, is the preparation of the application material. Before even starting to work on the application material, the secret is to read the job description very carefully. It may seem obvious, but understanding the advertised responsibilities expected for the specific position helps with the application package preparation. In an academic job search, a *search committee* is responsible for selecting the curriculums of interest and for performing the interviews. From all the applicants, usually only three or four candidates are interviewed. This fact shows how important is to prepare competitive and professional application materials. The documentation required in an academic job search may vary depending on: (1) the position, (2) the committee selection criteria, and (3) the type of appointment. In general, the committee asks for the following documents to be submitted:

- Cover letter
- Teaching statement and philosophy
- Statement of past and future research

- Curriculum Vitae
- Reference letters

Cover Letter

The cover letter should be short and specific. It is the first document that the search committee reads. From the cover letter, the committee develops a first impression of the candidate; they look for information about how well the candidate will fit the position and what contributions the candidate would make to the department, program, and curriculum. Therefore, the cover letter should state specific capabilities of the candidate by addressing the requirements of the job description. It is important to be clear, specific, and concise. The cover letter should start with a strong, but short, paragraph (no more than two or three sentences) stating the purpose of the letter. The next paragraph should contain a brief explanation of why the candidate thinks he or she is the perfect match for the position. A couple of sentences describing how the university is going to benefit from the candidate's expertise should be also included. The last paragraph should confirm the candidate's interest for the position and his/her willingness to meet with the committee to discuss further details. Finally, an acknowledgment to the committee for taking their time to read the application material is usually included in the cover letter. Some tips for writing a successful cover letter are:

- No spelling or typing errors
- Address it to the committee chair
- Write it in your own words so that it sounds like you—not like something out of a book
- Show that you know what the position is about and that you are the perfect fit for it
- Use terms and phrases that are meaningful to the committee

The committee won't want to spend more than two minutes reading a cover letter. The objective of the cover letter is to get the search committee excited about the candidate and encourage them to read the more detailed description of the candidate's capabilities. If the committee doesn't get excited with the cover letter, then they will most likely approach the rest of the material in a very conservative manner without many expectations. With first impressions being very important, the cover letter is a powerful tool that the candidate should take advantage of. A good and professional cover letter will significantly increase the chances of getting a job. If a weak cover letter is submitted, the search committee won't be able to find in this first document the attributes required in the job description and therefore, the risk of putting the application package aside will be higher. In summary, spending time and effort in writing a competitive cover letter is an important start to the application process.

Teaching Statement and Philosophy

Depending on the type of profile that the academic institution is looking for, a teaching statement and philosophy might be required as part of the application. As an aspiring professional in academics, it is important to start thinking about a personal teaching statement. A teaching statement describes the *why* and *how* of a personal teaching profile. Basically, the answers to the questions of *why* and *how* specific teaching styles are followed need to be addressed in this document. This includes a description of the outline and introductory message that is delivered to the students, the desired outcomes from the course, and the evaluation of those outcomes; the type of teacher–student relationship and the specific tools used to follow that teaching style should also be described. The goal of a good teacher should be to encourage students to learn and maximize the outcomes of the learning experience. Teachers might approach this objective with different styles. There is not a "good" or "bad" style. The important thing is to keep the main objective in mind and make sure that it is successfully met. Therefore, when preparing the teaching philosophy, it is key to be honest in the description of the teaching style, so that the search committee can establish if there is a fit between the candidate's and the department's style and philosophy. Additional guidelines about teaching strategies and philosophy can be found in Mack-Kirschner (2005), Achacoso and Svinicki (2004), Seldin (2004), Bullock and Hawk (2001), Hutchings (1998), and Constantino and De Lorenzo (2006).

Statement of Past and Future Research

This section of the application package needs to address the candidate's research experience and their plans on establishing an independent research program. It is important to make a distinction between this piece of documentation and the curriculum vitae, especially when describing past research experience. The statement of past research should describe the accomplishments obtained during the different stages of the candidate's career. That is, what was the candidate's role as a MS student, as a PhD student, and/or as a Research Assistant? If industry-related research or internships were performed, the personal experience in this position and why an academic career is being pursued versus an industry one should be explained. This statement is the opportunity to tell the search committee about the candidate's expertise as an independent researcher, as a critical thinker, and his/her potential to establish an independent research program. To demonstrate that the candidate is an independent researcher, he/she must show, through his/her accomplishments that he/she is able to successfully develop a research program, from the generation of novel ideas, the experimental design, and the implementation of the experiments to management of students or personnel and data interpretation and dissemination. It is important to note here that being an independent researcher does not mean that the person works isolated in his/her office without any interaction with colleagues. Researchers can build an independent program while still establishing successful collaborations either in the same department and/or institution or with colleagues from other institutions. To make sure that the researcher keeps his/her identity in these collaborations, his/her specific area of expertise needs to be clearly established in the collaborative work. As a matter of fact, showing interest in establishing collaborations with colleagues either from the department, the university and/or other academic institutions in the country is desired by the committee. They need to know that the candidate is willing to share his/her expertise with other members of the scientific community. Information that the search committee likes to see of past research includes the candidate's experience on: (1) supervising students and/or employees, (2) writing papers, (3) writing grants, (4) critical thinking, and (5) generation of novel ideas.

The statement for future research should include areas of interest to the candidate and how they will fit the department mission and curriculum. Search committees like to know that the candidate is willing to participate and become part of the academic community and not work isolated in his/her office or laboratory. Therefore, including in this statement a description of possible interactions and collaborations with other members of the department and of the university would be advisable. Finally, a long-term goal for the proposed research should be included in this section. Search committee members need to understand what the impact of the proposed research is and to be confident that a tangible research plan is described in the application material. The search committee members are looking for areas that are common between the candidate's application and the university's mission statement and objectives.

In summary, this statement should be a tool to understand the professional trajectory that the candidate envisions for his/her future. The candidate must clearly describe the type of researcher that he/she wants to become and the areas of interest that he/she will like to gain expertise in. This will enable the committee members to visualize the big picture of the candidate's future based on his/her previous experience.

Curriculum Vitae (CV)

Preparing a CV is a time-consuming activity. This document describes in detail the candidate's educational and employment records, including training, teaching, and research expertise. The search committee will find in the CV details about dates, places, job experiences, and performance. CVs should be prepared over time and reviewed and updated continuously; they should be concise and presented in an orderly format. Individuals reading the CV should be able to find specific information fast and efficiently. In general, CVs are organized in several sections. Some of the sections include: (1) personal and contact information, (2) education, (3) work experience, (4) awards, (5) research experience, (6) teaching experience, (7) professional activities, and (8) extracurricular activities. Each section (i.e., 2–8) must contain a list of accomplishments, including dates and places. CVs can have different styles; however, a short and easy-to-read CV is the suggested format that many search committees would find appropriate. The CV should be a document the interviewers can reference when they have a question. Therefore, to facilitate the reading of this material, including section headings and bulleted points is a good idea.

The most important aspect to consider when putting a CV together is to analyze the job offer description. The CV should be tailored to a specific job search by highlighting the activities that are most relevant to the job description. For example, if the job responsibilities will involve 85% research effort, then the research experience should be one of the first sections in the CV, with a detailed list of publications, scholarships, and grants received. Although CVs are developed around a standard template, making changes to fit a specific job description is a good strategy to follow.

It is worthy to make a note in this section regarding the difference between a CV and a resume. Not many people understand what the differences are and tend to confound them. The word resume is French and means "summary." Resumes are generally used in business and nonprofit settings and are usually 1–2 pages long; while CVs, which are usually used in academia, can run for pages and pages. In general, CVs present more details regarding the subject's expertise. For example, people would include in the CV the name and title of their PhD and/or MS thesis and advisor. Academia is a small world, and it is likely that a prospective employer will have heard of a given specialist in the candidate's field. In addition to the usual catalog of degrees and job histories, CVs contain many more categories of information than a resume. For example, experience may be divided between headings for teaching and professional; education may be divided between degrees and advanced training; publications may be divided between articles, conference presentations, and unpublished works. When applying to an academic job, it is very likely that the search committee will ask for a CV. Spend time preparing the CV and include as many significant details as possible. Guidelines to writing a CV along with typical examples can be found in Biegeleisen (1991).

Reference Letters

Usually academic job applications require three or four reference letters. These are letters of recommendation provided by someone who is familiar with the candidate's job and expertise. In general, reference letters are written by former supervisors and/or collaborators and they should describe the candidate's performance as a student, employee, and/or professional, and the candidate's potential to fulfill the specific job requirements. Reference letters are an additional source of information for the search committee providing an insight on the candidate's impact on colleagues' opinions.

In conclusion, the application package must reflect not only the candidate's capabilities, but also his/her enthusiasm for research, teaching, and academia in general. When preparing your application material don't be afraid of looking for advice. Books and Internet resources are good places to start. However, once you have the first draft of your application, you can ask for advice from colleagues and/or major professors. Sometimes, after working for a long period of time on a document, mistakes and errors are not noticeable anymore to the writer. Therefore, having a third-party opinion on the application material is a good idea to catch up on those. The first thing to look for is grammar mistakes and spelling errors. Make sure that your application material is free of these pitfalls. Content organization is also something to look for. Ask your reviewers to pay special attention to the dynamics of the application material content. Occasionally, it is difficult to put in paper what is in the writer's mind. Writers tend to avoid statements that are "obvious" to him/her but not necessary to the reader. Even though including specific and scientific terms related to the candidate's expertise in the research description is a common practice, keeping a simple style is always a good idea. Remember that it is likely that most of the committee members are not experts in your area of research, and therefore, they need to understand the content of the application material and the importance of your research. Therefore, when reviewing your application material, include at least one or two people that are complete strangers to your research area and test the simplicity of your statements. If these external reviewers can understand your research interests and how important they are to help the community, then you'll have an appropriate application material.

What to Expect in an Onsite Interview

Prior to an onsite interview, the application goes through a rigorous process. The committee's first step is to narrow the selection by setting aside those candidates who did not fulfill the requirements stated in the advertisement. From the rest of the applicants, each committee member grades the capabilities of each candidate separately. General characteristics such as area of expertise and type of degree obtained are some of the criteria that the committee members look at. Afterward, in a round-table discussion, the candidates to be interviewed are selected by consensus. Usually, the search committee selects three or four outstanding candidates to interview.

Academic interviews are exhausting and very stressful. The interviews are usually 2 days long. In general, the interview date is set a couple of weeks after the candidate is notified that he or she has been selected. These weeks should be used by the candidate to prepare for this stressful experience. Sometimes, the interview date is actually several months after the application deadline; therefore, it is wise to review the job description and the application package. An additional preparation tool for the interview is to study the committee members' profiles. Researching their areas of expertise and their recent accomplishments is a good strategy to follow. In this way, the candidate can tailor his/her answers during the interview according to the expertise found in the audience (i.e., his/her search committee). Depending on the job description the interview may consist of:

- Interview with the search committee members
- Interview with the department head
- Interview with the university's authorities such as Dean of the College, Provost, etc.
- Interviews with individual faculty members or small groups of faculty
- Interviews with student panels
- Research seminar
- Teaching seminar

All of these activities are supplemented by extracurricular activities such as lunch and dinner with different members of the academic community. The purpose of these activities is to provide a more relaxed environment to the candidate and indirectly evaluate if he or she will be a good match with the university's community.

Interview with the Search Committee Members

The interview with the search committee may last from 30 minutes to 2 hours depending on the committee and on the dynamics of the interview. The search committee usually consists of three or four faculty members from the department and one external member (from another department within the same academic institution). During this section of the interview, the committee members will ask specific questions about the candidate's experience as related to the job description and also on the contributions the candidate plans to bring to the university. The candidate has to demonstrate his/her capabilities and how he/she meets the requirements for that specific job. This is a crucial interview. This is the candidate's opportunity to answer all the questions that might have arisen from the submitted material. The candidate must keep in mind that other candidates with the same or better qualifications than his/hers are being interviewed as well; therefore, confidence and professionalism are attributes that the search committee usually looks for in a standout candidate. Ultimately, the candidate has to convince the search committee that he/she is the best choice for the position. Honesty and transparency are always good qualities to depict. Getting a job based on unreal qualifications will result in a waste of time and effort for both the candidate and the university. Specific advice for a successful interview will be discussed later.

Interview with the Department Head

When interviewing with the Department Head or Chair, the conversation will be focused on the department's activities and on the candidate's role in the department as a new faculty member. This is the right opportunity for the candidate to ask questions about the department and the work atmosphere. Questions that weren't answered either in the job description or on the university's Web page are appropriate at this stage of the interview process. This is the candidate's opportunity to evaluate if the institution matches his/her expectations. A perfect job is a two-way relationship between the employee and the employer. The employee has to make a significant contribution to the employer and vice versa; the employee should not feel that they are not benefiting from the position. Therefore, it is good to demonstrate to the search committee and department head the expectation of some type of benefit package from the university. An important point that might be discussed in this section of the interview process is the type of appointment being offered by the institution. Tenure track positions and 9-month appointments should be discussed and evaluated by the candidate. Procedures, policies, and expectations to achieve tenure and promotion are some topics that can be clarified by the department head.

Interview with the University's Authorities

This interview is very similar to the department head interview. In general, university's authorities are not going to be very interested in specifics about the candidate's accomplishments and/or about the specifics of his/her research and teaching strategy. However, they might be very interested in the candidate's strategy to develop an independent internationally recognized area of research, the candidate's experience in grant writing, and the rate of success in both areas. As mentioned in the previous paragraph, this is the perfect opportunity for the candidate to get answers about the university, its goals and mission, and its national recognition.

Interviews with Individual Faculty Members or Small Groups of Faculty

It is also very common to meet with some faculty members either individually or in a group. This is a great opportunity to discuss the research interests of each faculty member and evaluate the areas where future collaborations can be established. In general, faculty members have access to the candidate's CV and expertise so they can ask questions and evaluate how well the candidate would fit the department. Do not underestimate this part of the interview process, do the homework and research about the faculty members in the department. Knowing the specific areas of expertise in advance will give the candidate a good background to develop a good conversation and discussion about common research interests and future collaborations. Faculty members like to see a candidate that is active and energetic. Even though the candidate should let the faculty lead the interview, I strongly encourage the candidate to be active in asking questions to the faculty. Expressing interest in the different areas of research ongoing in the department will result in a very good impression and will open the door for a myriad of future collaborations.

Interviews with Student Panels

Students are a very important part of the academic community. Since most of the academic appointments include a significant teaching component, students have the right to draw an opinion on the future candidate's capabilities as a teacher. Student panels might include both graduate and undergraduate students. They might be interested in the candidate's teaching strategy and philosophy and also on his/her interaction with students as a mentor and supervisor. Students will usually look for a person that they can relate to, who is accessible, and willing to meet students' academic needs. When interviewing with students, just remember how your student life was. Your students days are probably not very far away in time and you, better than anybody else, know what a student expects from a faculty member. Going to these interviews in an open and relaxed attitude is a good strategy that will make students feel comfortable to ask questions and to establish a dynamic interaction.

Research Seminar

The objective of the research seminar, usually 50 minutes long, is for the candidate to demonstrate to the audience his/her achievements in research. The audience typically consists of the search committee members, the department head, employees, and students from the department and the university in general. The candidate's goal is to present to the university community a brief description of his/her area of expertise and the importance and outcomes of his/her research. In addition, future plans for establishing an independent area of research and the benefits that the university and the department will gain from hiring the candidate should be emphasized. However, the candidate must remember that many people in the audience might not be entirely familiar with his/her specific area of research since they might work in a completely different field. Therefore, maintaining a simple scientific language is a good strategy. The audience must be able to understand the importance of the research being described and encouraged to ask questions, which permits an evaluation of the candidate's ability to solve problematic situations. Candidates should not underestimate the importance of the research seminar. No matter how much experience the candidate has in public speaking, practicing the presentation beforehand is a good exercise that will help avoid last minute stress and confusion. Practice, practice, practice, and more practice is the secret to a successful presentation.

Teaching Seminar

When teaching responsibilities are detailed in the job description, a teaching seminar is normally included in the interview schedule. This is an opportunity for the candidate to demonstrate that he/she is an accomplished teacher. In general, 2 or 3 days before the interview date, the search committee sends the candidate a specific topic to develop for the lecture. The seminar topic is typically related to the teaching responsibilities stated in the job description. The search committee expects to see a teacher who knows the subject, encourages students to learn, and who has good interaction with the audience.

This is the candidate's opportunity to put his/her teaching philosophy and style into practice. Most academic appointments have an important teaching component. If this is the case, then the candidate's teaching performance during the seminar will be a determining factor in the search committee's requirements for choosing the right candidate.

Academic Interview Skills and Strategies

During the interview, it is very important for the candidate to have the right attitude. Honesty, professionalism, politeness, and self-confidence are attributes that the candidate must display during the interview. Candidates should plan for the interview in advance and as much as they can. The first step in preparing for a perfect interview is to research as much about the position offered as possible. Some information to research might include: (1) who was previously in the faculty position, (2) why did he/she leave the position at the university, (3) areas of expertise of the department faculty, and (4) the university's strongest areas of research, mission statement, and objectives. This information will provide the candidate with significant background and a good prediction of the interview dynamics. In addition, researching about the aforementioned topics provides the candidate with the tools to tailor his/her answers toward the areas of interest of the department and the university in general. Candidates are strongly encouraged to ask questions to the search committee since it shows his/her interest in the institution. The interview is the first impression that the search committee has of the candidate, and the decision to hire depends largely on the personal interview. Search committee members are part of the academic community, so they will be the candidate's future colleagues, and therefore, if the personal interaction is not good, the chances of choosing that candidate are very low. The candidate needs to demonstrate that he/she is an accessible and an easygoing person, whom people can relate to or at least see themselves working with in the future.

During the interview the candidate needs to watch his/her professional and personal behavior. Some recommended behaviors are:

- Personal appearance:
  - Hands—clean with a moderate nail length
  - Hair—should be done conservatively (no unnatural colors or styles)
  - Scent—remember too much cologne/perfume can be as off-putting as body odor
  - Breath—have a breath mint on hand

- General behavior:
  - Go to the restroom before the interview
  - Get a good night's rest
  - Be on time
  - Give a firm handshake
  - Wait to sit down until asked
  - Have good posture
  - Use gestures appropriate and sparingly
  - Smile
- About the interview:
  - Look the interviewer in the eye
  - Speak confidently and clearly
  - Get the interviewers' names right and use them
  - Study and research the academic institution in advance
  - Take notes (optional)
  - Have prepared questions
  - Let the interviewer set the pace and decide when the interview is over
  - Ask when you will hear from him/her
  - Follow-up (send a Thank you note)
  - Have extra copies of your resume and references

In the same context, some less-recommended behaviors are detailed in the following list:

- Personal appearance:
  - Don't wear an overcoat or boots
  - Don't wear a lot of jewelry
  - Don't have anything in your mouth
  - Don't carry an oversized handbag
- General behavior:
  - Don't lean on elbows or on interviewer's desk
  - Don't show nervousness (adjust clothes, hair, jewelry)
  - Don't swivel in your chair
  - Don't be late
- About the interview:
  - Don't go overboard with "Sir" or "Ma'am"
  - Don't call interviewer by first name
  - Don't give one- and two-word answers

- ○ Don't hog the conversation or interrupt
- ○ Don't use profanity or slang (even if a committee member does)
- ○ Don't chatter while interviewer is reviewing resume
- ○ Don't overstate qualifications
- ○ Don't criticize present employer or get angry
- ○ Don't feel obligated to answer personal questions, instead kindly ask what the question has to do with the job
- ○ Don't look at your watch
- ○ Don't ask "Will I get the job?"
- ○ Don't talk about salary unless the interviewer does
- ○ Don't hide anything

It is important to note that following this list of suggestions will increase the candidate's chance of getting the job. However, if any one of the nonrecommended behaviors is used, this will not necessarily disqualify the candidate.

As mentioned before, during the interview, the search committee expects questions from the candidate. Some questions that can be asked are:

- What are the most important skills for this job?
- Are there any job responsibilities that haven't been stated in the job description?
- How would my performance be evaluated?
- How would I be supervised?
- What is the last person who had this job doing? Where is he/she now?
- How long will it take to make a hiring decision?
- What does the university or department consider the five most important duties of this position?

In the same context, the interviewers might ask the candidate some questions that he/she should be prepared to answer. Some of these questions may include:

- What are your long- and short-term goals and objectives? When and why did you establish these goals and how are you preparing yourself to achieve them?
- What do you see yourself doing 5 years from now?
- What are the most important rewards you expect from your professional career?
- Why did you choose a career in academia?
- What do you consider to be your greatest strengths and weaknesses?
- Why did you select this university?

- Why did you choose your research interests?
- What contributions will you bring to the university?
- Why should we hire you?
- What qualifications do you have that will make you successful in this field?
- How do you measure the outcome of your research and teaching?
- Will you relocate?
- What have you learned from your past mistakes?

A last, but not least, skill that should be exercised is to follow up after the interview is finished. Follow-up letters should be directed to the search committee members and the people that the candidate met with. These letters should clearly state that the candidate is still interested in the position and that he/she had a good time during his/her visit to the academic institution.

Additional information about suggestions and typical questions asked during a job interview can be found in Biegeleisen (1991), Perlmutter Bloch (1987), Bermont (2004), Allen (1983), Hammond (1990), and Pettus (1979). Although these references discuss issues related to job interviews in general, these are usually applicable also to academic job interviews. In addition, details about academic interview procedures can be found in Vicker and Harriette (2006).

Negotiating an Offer

After all the candidates have been interviewed, the search committee will reach a decision and issue a recommendation letter to the department head. Procedures vary from institution to institution, and the candidate can ask about the specifics of this part of the search process during his/her interview. In general, the procedure consists of a letter from the search committee to the head of the department ranking the best three candidates. The department head, after approval from the dean of the college, will usually contact the first candidate in the list. When the candidate is contacted by the department head, a time is scheduled to discuss details about the job offer. The offer is not fixed at this time; it is a negotiation between the department head and the candidate. This negotiation can be done by phone, email, or mail. Negotiating the offer and benefits is a common practice in academic job offers; in fact, department heads expect some type of negotiation from the candidate. Benefits such as salary, moving expenses, and housing can be negotiated. Negotiating is not an easy task, though. The candidate must show confidence and must know exactly what he/she wants to get from this job. Details on specific items that can be negotiated is salary.

Salary

Negotiating the salary is a very common practice. The candidate needs to prepare for this type of negotiation. Researching the average salary range in the type of job that the candidate is being offered is recommended. Location, expertise, and university prestige are some factors that need to be taken into account when evaluating the expected salary range. The candidate must also keep in mind a salary range that he/she would like to receive. If the salary offered is below that range, then the candidate will most likely negotiate the amount. If, however, the offer is above that range, then the candidate should save negotiations efforts for other issues. If the amount offered falls within the expected range, then the candidate might choose to negotiate for a higher salary or accept the amount offered (this probably depends on the other benefits offered).

Appointment

Another important topic to negotiate is the type of appointment being offered. The tendency these days is to hire new faculty for a 9 month appointment, meaning that faculty members are paid on an academic year basis (September–May). This strategy has some advantages and some disadvantages. An advantage is that faculty can put those extra summer months salary into grant applications and, depending on their funding success a significant increase in the salary can be expected. However, a big disadvantage comes along with this issue. Grant money is not easy to get, especially federal grants, and therefore a new faculty member may not see a salary increase for a few years down the road.

Graduate Students

The productivity of an assistant professor during the first couple of years is very important. Their performance as a researcher and as a teacher will be exhaustively evaluated by colleagues and by university authorities. Both research and teaching outcomes are measured, among other factors, by the number of research papers published and the number of graduate students recruited. Therefore, the ability to obtain graduate students is crucial for an assistant professor. A very good idea is to negotiate some monies to fund one or two graduate students for at least the first couple of years. This will provide the new assistant professor financial security enabling focus on research and generation of preliminary data for grant submissions. Choosing the right graduate student is also an important issue that should be considered, but this topic exceeds the scope of this chapter.

Startup Money

Startup monies are funds provided by the department to help the new faculty to establish his/her research. These funds can be in the form of a monetary amount, which can be used for general expenses such as supplies, equipment, students, etc. However, sometimes startup monies are available only to purchase equipment. Both strategies are good and the candidate's choice will depend on his/her research interests and short-term goals. Before negotiating the use of startup monies, it is a good idea to check on the facilities the university already has, either in the department or on-campus. If the university already has the type of equipment needed to start the research program, a significant amount of money won't be needed to buy equipment. If this is the case, startup monies could be used to fund graduate students.

Moving Expenses

If the candidate is relocating, negotiating moving expenses is a good idea. Moving can be expensive and time consuming. Most of the time, it is not difficult to get the employer to pay for moving expenses or at least a part; sometimes, they set a maximum limit on the amount of money they will reimburse. Paid moving expenses is an important benefit, especially if the candidate wants to buy or rent a place to live and needs to visit before the appointment starts. In this case, the candidate might want to stay in a hotel for a couple of days to get to know the city, the different neighborhoods, and to make up his/her mind on where the best option is.

These are the most common negotiated items for an academic position. However, depending on the personal situation of the candidate, some other issues might be negotiated. For example, if the candidate is not a permanent resident of the country, he/she might also want to negotiate the reimbursement for expenses related to a permanent residency application.

Summary

Looking for an academic job is very exciting; from the moment the job search is posted to the interview. Every single step in the search results in a handful of rewarding experiences. Candidates usually learn not only about their professional and personal ambitions, but also about communicating them to others. Although looking for a job might be very frustrating, it is worth putting time and effort into the process. The most important thing to remember is to be honest during the entire job application process. If the job application wasn't successful, don't feel discouraged, another opportunity will arise. Finding the right candidate for a specific job is a difficult task. Considerable time needs to be spent in this endeavor to avoid wasting the candidate's and the university's time and effort. The ultimate goal of a job search is to ensure that both the candidate's and the university's expectations are met.

References

Achacoso, M.V. and Svinicki, M.D. 2004. *Alternative Strategies for Evaluating Student Learning*. Jossey-Bass, San Francisco.

Allen, J.G. 1983. *How to Turn an Interview into a Job.* Simon and Schuster, NY.

Baez, B. and Centra, J.A. 1995. *Tenure, Promotion and Reappointment: Legal and Administrative Implications.* ASHE-ERIC-Higher Education Report No. 1. Published by the George Washington University, Washington DC.

Bataille, G.M. and Brown, B.E. 2006. *Faculty Career Paths: multiple routs to academic success and satisfaction*. American Council on Education/Praeger series on higher education. Westport, CT.

Bermont, T. 2004. *Ten Insider Secrets to a Winning Job Search.* Career Press. Franklin Lakes, NJ.

Biegeleisen, J.I. 1991. Job Resumes: How to Write Them, How to Present Them, Preparing for Interviews. Perigee. NY.

Bullok, A.A. and Hawk, P.P. 2001. *Developing a Teaching Portfolio: A Guide for Preservice and Practicing Teachers.* Merrill Prentice Hall, Columbus, OH.

Constantino, P.M. and De Lorenzo, M.N., 2006. *Developing a Professional Teaching Portfolio: A guide for success.* Second Edition. Pearson Ed., New York.

Hammond, B.R. 1990. *Winning the Job Interview Game: New Strategies for Getting Hired.* Liberty Hall Press, Blue Ridge Summit, PA.

Hutchings, P. 1998. The Course Portfolio: How Faculty Can examine their Teaching to Advance Practice and Improve Student Learning. American Association for Higher Education, Washington DC.

Mack-Kirschner, A. 2005. *Straight Talk for Today's Teacher: How to Teach so Students Learn.* Heinemann, Portsmouth, NH.

Perlmutter Bloch, D. 1987. *How to Have a Winning Job Interview.* VGM Career Horizons, Lincolnwood, IL.

Pettus, T.T. 1979. *One on One: Win the Interview, Win the Job.* Focus Press, Inc., NY

Seldin, P. 2004. *The Teaching Portfolio: A Practical Guide to Improved Performance and Promotion/Tenure Decisions.* Anker Publishing Company, Boston, Massachusetts.

Vicker, L.A. and Harriette, J.R. 2006. *The Complete Academic Search Manual: A systematic Approach to Successful and Inclusive Hiring*. Stylus Publishing, LLC, Sterling, Virginia.

Chapter 29
Getting Started in Your Academic Career

Kalmia E. Kniel

An academic career is certainly one of great fulfillment and pride. After completing graduate school and postdoctoral work, perhaps even spending some time working in industry, it is now time to set up a laboratory, interview graduate students, find funding for these graduate students, adjust to becoming more of a lab manager than a researcher, write several grant proposals, write a syllabus or two, review manuscripts, and write a few of your own (or at least edit your students'), and serve on a myriad of university and departmental committees. Working in academia is a challenging and rewarding career. Being surrounded by the energy of students is contagious, and as you will discover an assistant professor needs that energy. An academic career at any level is the epitome of multitasking; juggling research, teaching, and service activities.

As you attempt to settle in, pay close attention to your surroundings. A good piece of advice is to pay close attention to your peers. Take notice of how these faculty members interact with one another and with the institution as a whole. Collegiality and developing a positive work environment can foster creativity. Unfortunately, many academic work environments are competitive, brought about by reduced government funding; therefore, make attempts to have healthy competition and offer collegiality where possible. Notice how your colleagues and peers collaborate and create opportunities where your skills can be valued and utilized. Do not be afraid to seek guidance and communicate openly with your colleagues about where they publish their research and how they advise their graduate students.

Mentoring is important and is handled differently at each institution. For some, mentoring is noted as a crucial step in the development of a junior faculty member (Trotman, 2006). If possible, upon arrival in your

K.E. Kniel
University of Delaware, Newark, DE, USA

R.W. Hartel, C.P. Klawitter (eds.), *Careers in Food Science*,
DOI: 10.1007/978-0-387-77391-9_29, © Springer Science+Business Media, LLC 2008

new department, find a tenured faculty member who is willing to provide you with guidance and mentoring as needed throughout the next few years. A good individual to consult is the chair of your search committee. Some departments have a system setup for mentoring and while it is recognized as important by most, it is often forgotten; therefore, it may be up to you to ask questions and find someone willing to provide guidance and lend an ear when necessary. This is the most frequently cited criterion for success (Trotman, 2006). To make this relationship the most fruitful, the mentor should benefit in some manner from the new faculty member, for example, from the success of the new faculty member. The department chair can be active in this forum as well. The department chair can act as a "career sponsor" (Bataille and Brown, 2006). The mentor and/or chair can help design a framework for a checklist, which should contain 2, 4, and 5 year goals. This framework can assist in the development of perhaps the most important criteria for early-career success, a clear outlook for tenure and promotion (Trottman, 2006).

The Quest for Tenure

Is tenure synonymous with the Holy Grail? It is likely the unknown and indefinable, which makes tenure somewhat mysterious. Attempting to "demystify" this process may not be simple or plausible (Bataille and Brown, 2006). At the very least, a faculty handbook should be made available at the start of your quest through this process. Often colleagues will offer their dossiers as examples, and this is an excellent means of identifying early on what you should develop over the course of the next few years. Evaluation occurs in every career path and the quest for tenure occurs in every academic institution around the world. Luckily, the evaluation process should be ongoing to make the process fairer and somewhat less challenging, and perhaps even more survivable (Diamond, 2004). Typically, there are annual reviews within your department, as well as 2 and 4 year reviews of your developing dossier, which provide guidance for dossier structure. The goal of these departmental committee reviews is to identify strengths and weaknesses of the assistant professor so that he/she can build upon them prior to coming up for the tenure evaluation. The more information you have going into the process the better off you will be in the end. Be aware of the criteria; however, this may not be clear and certainly may not be given as numbers. The criteria are subjective and will be affected by the discipline of the reviewer. Be sure to explain aspects of your dossier thoroughly and justify why the documentation speaks of your excellence in your career.

Use your offer letter as a direction on how much time should be focused in research, scholarship, and teaching. This information can be useful in attempting to answer the question, "For what is tenure awarded?" which is asked by many and answered by few (Rice and Sorcinelli, 2002). It is essential to integrate scholarship into this balance, but what exactly is scholarship? In 1990, Ernest Boyer proposed that colleges and universities expand the traditional roles of research and teaching. The definition of scholarship includes several aspects (Boyer, 1990), including: the scholarship of discovery (original research), the scholarship of integration (synthesizing and reintegration of knowledge), the scholarship of application (professional practice), and the scholarship of teaching (transforming knowledge through teaching). For many food scientists, scholarship activities integrate teaching and outreach, along with aspects of research. It is important to be rewarded for this work, including what you may be very passionate about like engaging others in food science. Institutions are being increasingly pressured to reward faculty for this work (Rice and Sorcinelli, 2002). In food science, scholarship may include a myriad of activities including some directly related to applied research and others that involve recruitment and teaching workshops. An activity can be considered scholarly if it meets a variety of criteria (Diamond, 2002), including that requiring a high level of discipline-related knowledge, is conducted in a scholarly manner in terms of goals, preparation, and methodology; results are appropriately documented, the significance is beyond the individual context, and the process and the result are viewed by your peers as meritorious or significant. Ideally, multiple forms of scholarship should be rewarded. While institutions may broaden their horizons and encourage faculty to participate in these various forms of scholarship, it is likely that early-career faculty will need to devote the majority of their time to the scholarship of discovery in the form of research (Rice and Sorcinelli, 2002).

Preparing Your Dossier

After selecting your office furniture and arranging your desk, label a drawer in your file cabinet "Promotion and Tenure," and for the next five years, anytime you receive a thank you letter, publish a paper, get a request for an interview, etc., place proof of this interchange in that drawer (Kitto personal communication). The dossier serves as your organized collection of documentation. The role of this documentation is to provide support for your claim as a successful faculty member and food scientist. The documentation should speak to your collegiality, your effectiveness as an instructor, your effectiveness as a team member, collaborative work, your role as a researcher, and your success in teaching.

A first simple step is to keep your resume up to date. This will most likely be accomplished without thinking about it due to annual reviews and grant proposals, but make a conscious effort to update your resume as courses are taught, students graduate, and presentations are given. The dossier is an educational tool and will be used to educate your committee about your accomplishments (Diamond, 2004). The manner in which your dossier is assembled is critical as it will be evaluated by a committee composed of faculty outside of your discipline. Be sure to make information easy to find within the dossier. Obtain guidance from your mentor and department chair on how best to portray student evaluations and other data components of your dossier. Easy to view charts and graphs are helpful. Alongside the useful documentation, the dossier provides an opportunity to express your mission and visions on teaching, research, and scholarship. It is important to represent institution and departmental goals in these documents. Be weary of profound differences in disciplines that could influence or lead to miscommunication. To avoid these potential pitfalls, be sure to justify and explain the documentation provided in the dossier, including reasons for publishing in specific journals. Send some time on your teaching and research statements and revaluate these as time passes and learning continues.

Writing You First Grant Proposal

Obtaining funding is a crucial factor for a successful academic career. The old adage "try and try again" certainly pertains to grant writing, as funding is competitive. Prioritize and strategize your efforts. If possible in your field, apply for both large (federally funded) and smaller grants (local industry contracts, federal dollars, college or university money). Today the majority of funded grants take the multidisciplinary approach to solving problems. This provides the early-career faculty member opportunities to meet others interested in a similar field. It also provides an opportunity to demonstrate specific research talents that may add to another's grant proposal. It may be essential to seek funding in an area of research that is not your expertise or is an area in which you are growing. Step outside your comfort zone, but be cautious to ensure that you are able to organize all your research interests under one larger topic. This point will be crucial when you go up to tenure and are evaluated by your peers. In food science, perhaps more than in other sciences, faculty are forced to switch topics as problems arise in the areas of food safety or processing as research funding priorities (RFPs) change. This has certainly been true over the past three years. For example, proposals submitted for the United States Department of Agriculture National Research Initiative (USDA-NRI) funding program must meet the specific RFPs or they will not even be read by the reviewers. These RFPs are often more limited than they are broad in scope; however, this may change over time. This enhances the need for collaboration in these projects. Multidisciplinary and multi-institutional projects are more common now then they were 10 years ago. A way to interact with faculty from other universities is to join a multiregional HATCH project and your dean's office should provide guidance on how to do this.

There are some basics to writing a good grant proposal. The first is to read the request for funding applications (RFA) carefully. Develop ideas that fit within the program priorities. Consider your eligibility for various programs within this RFA, such as young investigator or seed programs and EPSCOR states (the Experimental Program to Stimulate Competitive Research) states. The later represents a family of federal state science and technology programs administered within the USDA and six other federal agencies. EPSCOR was established to identify and develop a state's academic science and technology resources in a way that it will support the state. A state can participate in EPSCOR based on the amount of funding, and it is certainly advisable to determine if your state is an EPSCOR state before looking at funding agencies.

It may be helpful to read a successful proposal from a colleague and follow this format for your first proposal. Ask a colleague in your research field to review your proposal for clarity and logic. The proposal will be read by at least three peers during the review process, and they may likely be outside of your research field. A successful proposal can excite the reviewers while being easy to read and understand. Have a clear rationale and clear objectives within your proposal. Do not make your reviewers search for answers, but rather provide them with adequate background on the experiments and justification for the work. Include a discussion of expected outcomes. Include clearly defined, integrated elements in your proposal if that is necessary for the program; for example, in the USDA integrated grants, be sure to follow all submission rules and clearly state the qualifications of each investigator. Spend adequate time on your summary statement, as this may be the only part read by the decision maker (CSREES, 2007). It is important that your enthusiasm for your work be evident in your writing. As you would tell your students, do not let typographical errors sway your reviewer form recommending your proposal.

It may be difficult to obtain funding in the beginning as a sole-principle investigator or as the lead and while times have changed, your departmental promotion and tenure committee may not comprehend these changes. Be sure to apply for funding from a variety of sources. Initiate contracts with private or corporate businesses and develop contacts that will help you gain experience in your new field or help you meet other collaborators. Smaller projects (1–2 years) can often develop into larger federally funded projects (3–5 years). Your institution will likely have resources as well. Contact the dean's office for funding opportunities. The dean of research is an excellent source for collaborations within your college or university. Talk to your colleagues about important issues in your own state. There are likely opportunities for funding to solve these issues. If possible, obtain preliminary data. This type of data is often less fundamental for seed proposals, but in general the ability to do the work will impact the success of the proposal.

Teaching Your First Class

Regardless of your teaching experience in graduate school, teaching your first college class will no doubt be a mixture of excitement, optimism, and anxiety. You may have accepted your position agreeing to teach a specific course or you may have the opportunity to develop your own course. Surprisingly, teaching is often not the primary reason one chooses to become a college professor (Fink, 1984). Despite this fact, teaching is of the utmost importance as the college was built for the students and everyone acknowledges this fact, including the department chair, college dean, provost, and university president. All institutions have a strong commitment to faculty development related to teaching. For example, in 1975, the faculty senate at the University of Delaware established the Center for Teaching Effectiveness (CTE) to recognize that Delaware is a teaching-oriented university with a strong academic reputation. Upon arrival at your institution, seek out these types of organizations to receive instructional support and creativity. While traditional lectures still have their place, optimal learning includes engaging students in creative manners to increase their understanding. This includes implementing programs and activities that enrich and improve teaching and learning. For example, group activities develop skills in leadership, cooperation, deductive reasoning, and compromise. Do not be afraid to seek guidance in the development of these activities. There is no reason to reinvent the wheel and by seeking out assistance from the Office of Undergraduate Studies, it is possible to gain access to a wide variety of teaching styles. For example, the CTE coordinates teaching conferences, workshops, colloquia, and publishes and disseminates materials on instructional practices and student learning. As mentioned above, teaching is a significant component of the dossier and it is important to include examples of syllabi and activities, course assessment, and student evaluations.

When developing your course and your teaching style learn by observing the classes and syllabi of your colleagues. Define your audience, undergraduate or graduate students. The students of today have had access to more information than any other cohort in society (Gardiner, 1994). This cohort has also grown up with "helicopter parents" and instant gratification. These both greatly impact the ways in which students learn and their expectations from class. As a teacher, it is important to display knowledge and subsequent learning in a variety of ways considering the different learning styles of the students. Advice for maintaining students' interest and preparing your teaching can be summarized in the following six themes (Fink, 1984):

1) Being as prepared as possible with course lectures and aids before you teach will help ensure that the course runs smoothly throughout the semester. It is important that teachers be prepared ahead of time for class and try not to wait until the last minute until you are comfortable with your course.
2) Plan to work hard as a teacher. As stated above, your students are knowledgeable and you must work to be steps ahead of them. One of the fun parts of teaching is the excitement of a learning student and the genuine surprise on the part of the teacher concerning the interesting questions that arise.
3) Be flexible but firm and consistent. Being well prepared will in turn allow you to be more flexible. This point applies to grading as well as other teaching applications.
4) Get to know your students, watch their reactions, and listen to them. Again learning from your students will make the process more rewarding.
5) Realize that you will make mistakes and do not get depressed or attempt to overdo it. Both humor and self-confidence are essential.
6) Learn about yourself as a teacher and constantly evaluate yourself. Try to sit in on others' classes to learn from their teaching styles.

In addition to teaching assistance, organizations like CTE offer instructional improvement grants in specific areas of teaching and learning. Often these grants are to advance the institution's teaching mission. Becoming a fellow of an organization like CTE or obtaining an instructional grant can add greatly to your dossier. Instructional grants may be available to enhance specific courses or core goals across the curriculum. Teaching collaborations can produce lucrative grants as well. Perhaps a course can benefit from experts in multiple disciplines. For example, food science, mathematics, and chemistry faculty can all work together as well as faculty studying food science, microbiology, and biology. Instructional grants may promote the use of technology in the classroom. For example, a grant award may be paid in time with a technology expert in the development of computer-based educational games, wikis, or other means of assessment. It is always best to use technology where it can be useful and not just to use technology; however, be aware that your students are used to learning with integrated technology. Applications of technology should promote student involvement and learning (Treuer and Belote, 1997). For example, electronic games can be used to reinforce concepts for visual learners or to entice students to critically evaluate a case study in multiple steps.

Managing a Laboratory and Graduate Students

This is perhaps the greatest change in a young faculty member's life. The transition from researcher to laboratory manager occurs quickly. There are many items to consider including equipment, selecting graduate students, obtaining graduate student stipends, and participating in undergraduate research training. The latter requires upfront investment time, but is often quite rewarding for both student and advisor in the end. By having undergraduate students work on smaller, more manageable aspects of projects you will have a greater chance for success.

In arranging your laboratory and initiating graduate students and undergraduate students, make lab safety a priority. Take the necessary precautions. Academic labs work differently from industry labs as they have a lack of structure and academic scientists are free to explore without many institutional or marketing constraints; however, there is a risk to this informality in that people with minimal training may do dangerous work (Austin, 2006). Safety is taken seriously at most institutions, but do make this a priority in your laboratory. Start by training your personnel adequately from the start, including students. Also, initiate biological and chemical hygiene plans and records as you move into your lab rather than going back and doing it once the lab is filled. This will also make it more feasible for you to work on your course work and write grants when you know the lab is in safe hands.

In negotiating your contract, you may have been given startup funds to hire a research associate or graduate student. It is best to know this type of information as soon as possible so that you can prepare in advance. If the department has teaching or research assistantships available, you should be given one of these as a new faculty member. Become active in the graduate program committee and gain access to the electronic database of graduate applicants. Familiarize yourself with what your institution or colleagues view as acceptable entrance scores. If possible interview and ask colleagues and other graduate students to interview potential graduate students. Trust your intuition when considering graduate students. If you can develop a good correspondence with a student, the chance is good that he/she will do well over the next two years. People can learn laboratory skills, but they will never learn to love science or learn the motivation it takes to complete a graduate degree, and you will be able to sense this within essays and in particular in an interview.

Academic and Professional Service

Extension and outreach are essential components of an academic career. Faculty members need not have an official extension component to perform service duties or serve on various committees within the department, college, or university. In fact, throughout the first four years of an academic career it is important to gain service experience on all levels. Academic service can take on many forms. Some examples of these include serving on search committees or curriculum committees at the department level, participating in AgDay or recruiting events for the College and being elected as a faculty senator for the university or contributing to university honors' programs. Service responsibilities are not created equally, but they are all important and necessary for the institution to function properly. Initiating your faculty career includes an automatic agreement to working in committees or perhaps even a committee on committees. In addition to research and teaching, these service responsibilities will be evaluated during the promotion and tenure process. Striving for a balance over the first five years is important, but may not be easily achievable depending on your position, discipline, and department. Often the department chair will make an honest attempt not to overload a new faculty member with committee work (Gelb, personal communication). It may be useful to observe the committees or at least find out details about the duties before agreeing to serve on one; however, if asked to serve, it is not easy to turn down the chair's request. Talk with your chair and colleagues about your interest in serving on committees and your hesitance about being too caught up in one that it takes time away from research or teaching (depending on your work load). As you become acquainted with more people within your institution you will likely be asked to participate on more committees. It may be useful early on to seek out the ones you are more passionate about, i.e. those concerning student life, general education goals, or curriculum changes. Serving on committee will help you learn how your institution runs, which in turn will help you develop into a better advisor and colleague on many levels.

Outside of academic service, professional service activities are rewarding and a duty of all faculty members. While they can take time away from research and teaching, these activities are an excellent means of meeting other professionals in academia, industry, and government. Professional service activities are those that relate specifically to your discipline and may be centered on one or two societies. In food science, the Institute of Food Technologists (IFT, www.ift.org) certainly has countless opportunities for service. Take advantage of the expertise from members of IFT and other societies. Performing professional service activities will broaden your networking circle and enhance opportunities for research and teaching collaborations.

If you choose not to participate in service activities at the academic or professional level, this will be surely be noted during you annual evaluation and 2 year review; therefore, it is important to be proactive about initiating these contacts from the start. It may be quite feasible to build upon contacts from graduate school in IFT or other associations. On easy first step is to locate the smaller divisions and local chapters within the larger associations. Leadership roles within these smaller divisions are an excellent means of spreading your name and becoming known in your field or perhaps even slightly outside of your field.

Documentation concerning service activities should reflect the importance of the activity. The weight given to these activities will be a function of your academic appointment (percent teaching, research, and extension) (Diamond, 2004). In saying this, if your faculty appointment is 60% research, a leadership role with the IFT will likely be weighed more heavily compared to your participation in a departmental committee in terms of research. The opposite could be true for a 60% teaching appointment depending on the division and on a positive note, several divisions exist for educational purposes, including the IFT Education Division. The significance of the activity and the impact on scholarly work should also be considered in terms of documentation and the description of the activity within the dossier. For service to be considered scholarly without question, it should require a high level of discipline expertise and move your field ahead and have applications beyond your own institution. It is likely that there will be many opportunities for academic service to impact your field.

Summary

An academic career is fulfilling and challenging in many ways. It is the only career where you are, in many senses, your own boss and you must have the vision and motivation to persevere. It will take a great deal of energy to accomplish so much in less than 6 years, but it is possible. Through observation, collaboration, and self-determination, you will find success in teaching, research, and service. It is difficult to find the delicate balance at first, and no doubt, this balance will be upset more than once during the quest for promotion and tenure; however, if you continue the course, the rewards are great. Working in a field you love and surrounded by the energy of students and ambitions of science is a brilliant experience.

References

Austin, J. 2006. Staying well: Safety in the lab. Science Careers. http://sciencecareers.sciencemag.org/career_development/previous_issues/articles/2006_08_04/special_feature_staying_well_safety_in_the_lab/(parent)/13199. Accessed October 15, 2007.

Bataille, G.M. and B.E. Brown. 2006. *Faculty Career Paths*. Westport, CT: Praeger Publishers.

Boyer, Ernest, L. 1990. *Scholarship Reconsidered: Priorities of the Professoriate*. Princeton, NJ: The Carnegie Foundation for the Advancement of Teaching.

Cooperative State Research, Education, and Extension Service (CSREES). General tips for grant writing success. www.csrees.usda.gov. Accessed November 1, 2007.

Diamond, R.M. 2002. The mission-driven faculty reward system. In R.M. Diamond (ed.), *Field Guide to Academic Leadership*. SanFncisco, CA: Jossey-Bass. 280–290.

Diamond, R.M. 2004. *Preparing for Promotion, Tenure, and Annuals Review*. Bolton, MA: Anker Publishing Co., Inc.

Fink, L.D. 1984. *The First Year of College Teaching*. San Francisco, CA: Jossey-Bass, Inc., Publishers. 103–105.

Gardiner, L.F. 1994. *Redesigning Higher Education*. Washington, D.C.: The George Washington University. 7–24.

Gelb, J. Department Chair, Animal and Food Sciences Department. University of Delaware, Newark, DE. Personal communication 2006.

Kitto, S. Professor, Department of Plant and Soil Sciences. University of Delaware, Newark, DE. Personal communication 2004.

Rice, R. Eugene, and M.D. Sorcinelli. 2002. Can the Tenure Process Be Improved? In R.P. Chair (ed.),*The Questions of Tenure*, . Cambridge, MA: Harvard. 101–124.

Treuer, P., and L. Belote. 1997. Current and Emerging Applications of technology to promote student involvement and learning. In C.M. Engstrom, and K.W. Kruger (eds.), *Using Technology to Promote Student Learning: Opportunities for Today and Tomorrow.* San Francisco, CA: Jossey-Bass Publishers.

Trotman, C.A. 2006. Five criteria for early-career faculty success. In G.M. Bataille, and B.E. Brown (eds.), *Faculty Career Paths*. Westport, CT: Praeger Publishers, 66–75.

Chapter 30
Faculty Expectations and Development: The Tenure Case

S. Suzanne Nielsen

Introduction

Professionals seeking careers in academia should understand the tenure process, and how to prepare successfully for the evaluations linked to the tenure decision. This chapter offers suggestions for persons pursuing tenure–track faculty positions in the discipline of food science. The first promotion process in academia (i.e., from assistant professor to associate professor) is typically linked to tenure consideration. The focus of this chapter is explaining tenure, tenure expectations, resources for guidance, how to manage the process, and how to prepare the tenure and promotion document. While most people are fearful of the promotion and tenure process, this fear and apprehension can be minimized by understanding the process and its expectations, and having good advice to follow to help ensure success.

Explaining Tenure

Tenure has been defined as "assurance of academic freedom and permanence of contract in the sense that it may be terminated by either party only for just or serious cause" (Murphy, 1985). While tenure does not assure one of lifetime employment, it is a lifetime assurance in higher education that one will receive due process (Diamantes, 2002). The basic functions of a tenure process within an academic institution are to protect and facilitate scholarship, evaluate its faculty, and retain its best faculty members.

In tenure considerations, peers at an academic institution review the promotion and tenure documents of junior faculty to recommend dismissal or retention. The intent in the first promotion and tenure consideration is to

---

S.S. Nielsen
Purdue University, West Lafayette, IN, USA

R.W. Hartel, C.P. Klawitter (eds.), *Careers in Food Science*,
DOI: 10.1007/978-0-387-77391-9_30, © Springer Science+Business Media, LLC 2008

determine if the junior faculty member can be successful in the next promotion, i.e., is his/her program sustainable to justify promotion from associate professor to professor. This consideration of promotion and tenure is important because it makes faculty stakeholders in their institution, making them citizens rather than subjects. While the promotion and tenure process can be challenging and stressful for junior faculty members, understanding the expectations and taking advantage of mentoring from colleagues and other resources available can reduce the stress level and better ensure success.

Based on an extensive review of legal history regarding tenure, Olswang et al. (2001) identified a number of things tenure does and does not do:

What Tenure Does

- Provide protection of academic freedom (intellectual expressions and inquiries)
- Provide for employment security through a conditional employment contract
- Facilitate the employment of highly qualified people in a highly competitive market

What Tenure Does Not Do

- Protect against termination
- Allow faculty to say anything they want in the classroom
- Protect against a salary reduction
- Allow faculty to research any topic they choose
- Guarantee adequate space or equipment for research
- Permit faculty unrestricted authorization to submit requests for external funds
- Allow faculty unrestricted speech at faculty meetings
- Protect faculty from being punished for public statements about an institution
- Give faculty ownership in everything developed in the course of employment
- Give faculty the right to engage in unlimited outside consulting
- Allow faculty to determine the courses they will teach
- Give faculty the exclusive right to determine the content of their courses

Tenure Expectations

The productivity and merit required of faculty by an institution at the time of tenure consideration speaks volumes about the institution's values and priorities (Diamond, 2002). Promotion and tenure guidelines and expectations are typically described in terms of the criteria to judge scholarly, professional, and creative work (Diamond, 2002). The Boyer–Rice model, regarding the four general forms of scholarship, is used on many campuses (Boyer, 1990; Diamond, 2002):

- Discovery: advancing knowledge through original research
- Integration: synthesizing and reintegrating knowledge to reveal new patterns and relationships
- Application: using new knowledge in professional practice
- Teaching: transforming knowledge

Within the world of food science, faculty are asked typically to demonstrate scholarship in some combination of the following areas of responsibility: research (discovery), teaching (learning), and extension (engagement/outreach/service). Hopefully at your institution, each of these mission areas is valued, and you are guided to excellent examples of scholarly activity in each area.

Diamond (2002) has suggested that an activity or work will be considered scholarly if it meets the following criteria:

- It required a high level of discipline-related expertise.
- It is conducted in a scholarly manner with clear goals, adequate preparation, and appropriate methodology.
- The work and its results are appropriately and effectively documented and disseminated. This reporting should include a reflective critique that addresses the significance of the work, the process that was used, and what was learned.
- It has significance beyond the individual context.
- It breaks new ground or is innovative.
- It can be replicated or elaborated on.
- The work—both process and product or result—is reviewed and judged to be meritorious and significant by a panel of one's peers.

Resources for Guidance

While there are textbooks available to assist faculty as they prepare for promotion and tenure review (Boice, 2000; Diamond, 1995; Gelmon and Agre-Kippenhan, 2002; National Education Association, 1994; Miller, 1987), the best initial advice for new faculty members is to get to know their own institution and department, seek out mentors, and become familiar with the tenure and promotion guidelines/policies of the institution.

Orientation Programs

To help you prepare for mentoring programs that will assist you with the promotion and tenure process, you need to know the various resources available in the department, college/school, and university that might be of benefit to your specific program. To learn about these resources, you need to take full advantage of orientation programs at all levels in the institution—university, college/school, and department. During the first year of a faculty position, it is easy to think that you do not have time to attend all these orientation sessions. However, these orientation programs are well worth the time and effort. You will not only learn about resources on campus, but you will also meet senior administrators and faculty making presentations (who can be help to you in the future), along with other junior faculty who have a great deal in common with you and may become collaborators and/or friends.

Learning About the Department

The faculty, staff, and department head/chair can be invaluable resources to you as a new faculty member. Each of these can help you be more successful, as described below.

While you may have met many of the faculty members in your unit during the interview process, it will be very helpful once you begin as a faculty member to learn more about their roles, responsibilities, expertise, and resources. Take the time to visit with each faculty member and arrange to visit his/her laboratory, etc. This will help you learn what equipment and other resources each faculty member has, since you may need access to these later. Also, work to learn from faculty about their leadership responsibilities in the department, such as directors of centers, chair of graduate committee, chair of undergraduate committee, etc. Learning about their roles will help you understand how things work in the department, and how your responsibilities/activities might be connected to these subsets of the department (e.g., as you take on graduate students). As you visit with the various faculty members in your department, get advice about whom else on campus you should meet, based on the nature of your research program, teaching responsibilities, and/or extension program. Follow up by arranging to meet these other faculty members, etc. on campus. This will help you identify resources outside the department and collaborators for future grant proposals, etc. As you proceed with developing your programs, courses, etc., do not hesitate to take advantage of the expertise of persons you have met, ask their advice, or ask to borrow/use equipment.

Learning about your department should not stop with the faculty members. As part of your orientation to the department, you will be well served by meeting with key staff members to learn about their roles. Examples would be staff members who help handle the graduate program, manage the business office, run the pilot plant, manage the computer network, schedule classes, etc. Taking the time to learn about what they do will make it easier for you to carry out your responsibilities and be successful. Having a good relationship with staff members can be invaluable.

As you work to learn about the department, take the time to learn something about the responsibilities of the department head/chair, and the expectations he/she has of faculty. While an important role of the department head/chair is to help junior faculty members be successful, he/she has many other responsibilities toward the overall goal of helping the entire department be successful. Discuss with the department head/chair what he/she can do to help you be successful, what the expectations are for you, how your performance is evaluated annually, and how you can be of help within the department. Meet at least monthly with the department/head chair during your first few months on the job, to let him/her know what progress you have made to establish your programs, and to get advice and further direction.

Mentoring

Most departments have either formal or informal mentoring programs for junior faculty. Ask your department head/chair about mentoring programs, and take full advantage of such programs. Do not be hesitant to ask senior faculty members to serve as mentors in a formal way, or to simply ask their advice. Seek out, as mentors, well-respected and successful senior faculty members, matching them to you based on areas of expertise and responsibility. If you have responsibilities in multiple areas (e.g., research, teaching, extension), identify at least one mentor in each of the areas. These mentors do not need to all be in your own department, but they need to understand the nature of your responsibilities. Also, it is important that you consider personalities (i.e., yours and theirs) in selecting mentors, and make sure you feel comfortable with the mentors. Meet with your mentors both regularly and as needed—both with an agenda and informally. It is particularly valuable to discuss major decisions with your mentors, such as whether to take on a particular job or how to approach a major problem. Especially with your mentors, do not be afraid that your questions will make you seem naïve or unknowledgeable. Mentors want to, and are expected to, be helpful and understanding. One responsibility of senior faculty is to help junior faculty become successful, and mentoring is a key component of this process.

University Promotion and Tenure Policy and Instructions

At least within the first year of starting a faculty position, you should obtain a copy of the formal promotion policy for your university and the instructions for the promotion document. This policy and instructions may be modified yearly with regard to specific requirements about the promotion document, so you should check the policy and instructions periodically. Especially as the time approaches of actually being considered for promotion and tenure, ensure that the document follows the correct format and includes all the required information.

University promotion and tenure policy and instruction documents typically include the following information:

- Basis and philosophy for promotion to different ranks (i.e., tenure–track instructor to assistant professor, assistant professor to associate professor, associate professor to professor)
- Timeline/time table; effective dates
- Pre-tenure review policy (e.g., third-year review)
- Roles of department head/chair and dean in promotion process
- Required forms
- Format
- Content of promotion document (i.e., what to include in general information teaching, research, and extension sections)
- Criteria/standards and criteria indicators
- Make up of department, area, and university promotion committees, and rules that apply
- Information about letters solicited from outside referees (external peer evaluation)
- Early promotion/tenure
- Stopping the tenure clock; basis for automatic and requested extensions
- Salary increases linked to promotions
- Negative decisions
- Appeal procedures

Managing the Process and Preparing Tenure Document

Begin in Graduate School

Graduate students possibly interested in an academic career are very well served by being assertive about involvement in projects with the major professor. Obtaining experience in supervising undergraduate students or other junior members in the research laboratory, and helping to write grant proposals, both can be very instructive. Opportunities to help review manuscripts and grant proposals can be very valuable. The discussion that accompanies review of manuscripts/journal articles in the setting of laboratory meetings or journal clubs can be especially helpful. Teaching experience gained by giving lectures and assisting with or supervising laboratory sessions can be valuable, whether or not you intend to pursue an academic career. If you do intend for such a career, then graduate school is a great opportunity to take a class on learning methodologies, and to learn about teaching techniques and methods from some of the best teachers in your department (Abbott and Sanders, 1991). If your institution has a "teaching academy" or center of excellence for teaching, whose members present workshops on teaching or serve as mentors, take advantage of these resources. These teaching experts can help you build a network that will be of value to you well into your career. Some universities have programs available to graduate students and postdoctoral research associates about preparing for academic careers. Take full advantage of these.

The First Year

Before addressing specific suggestions for research, teaching, and extension very early in your career, some questions are posed for consideration. If you were not asked some of the following questions during the interview process, then ask them of yourself as you begin the job: Four to five years from now, when you want to be in a position to be promoted and granted tenure, what will you have accomplished in terms of impact of your program? What will be the evidence of your scholarly activities? What will your peers think of your work, and what products of your work will they recognize? What resources will you have brought to the table to help support your program (traditional and nontraditional sources)? How will you have engaged students in your program? The eventual answers to these questions will provide the data need for the evidence-based system of promotion and tenure. Junior faculty must provide good evidence to senior faculty and administrators to achieve promotion. This evidence needs to show consistency in productivity and it need to "tell a story." Promotion is based on a "body of work," rather than an unconnected series or projects.

New faculty members in the first year need to get a good start on all aspects of their specific appointment—teaching, research, and/or service. Many believe that quality teaching should be a goal of all faculty members, since it is integral to teaching success, and applying the communication skills of teaching is important to research and service. Participation in teaching workshops and taking advantage of peer evaluation can be extremely valuable. Since teaching competency is often measured by student evaluations, paying close attention to student needs and being available to help them are important (Abbott and Sanders, 1991). In addition to quality teaching, demonstrating scholarship with regard to teaching is of increasing importance (e.g., publishing in *Journal of Food Science Education*), particularly for faculty with primary teaching appointments.

The long time required from beginning a research project to the published papers requires that new faculty get started immediately on research efforts. These efforts typically include setting up a laboratory and writing grant proposals to obtain funds to support graduate students, etc. Preliminary data often are needed prior to writing major grant proposals. Working with others who share an interest in a common project can speed the process, make work loads more manageable, make the process more stimulating, and enhance the quality of the project (Abbott and Sanders, 1991). The emphasis by granting agencies and academic institutions on interdisciplinary research makes such collaborations essential. Learning about the expertise and equipment of others on campus, as described above, enables the creation of such collaborations.

Faculty with major extension/service appointments must learn quickly about the extension system in the state, relevant state agencies, and the needs of relevant companies/institutions/individual, etc. in the state. Obtaining good mentors in this area may be of highest importance, since students during graduate school receive even less training in this area than in teaching or research. It is especially important to learn what metrics (e.g., impacts, scholarship) are important for an extension appointment. Faculty with major extension appointments often have secondary research appointments, typically with a focus on applied research to complement their extension appointment.

Funds obtained from grants are usually critical to success in the promotion and tenure process—to support not only research programs, and also for teaching and extensions programs. The first step is to learn about funding opportunities specific to your academic institution, within the state/region (e.g., industry, commodity groups), and nationally (e.g., USDA, NSF, NIH). Another step is to learn about the grant submission process and policies at your home institution. Your business office, department head/chair, mentors, and other faculty will be resources for this information. New faculty should take full advantage of small grant opportunities typically available on campuses, sometimes targeted especially to junior faculty. New faculty also would be wise to collaborate with colleagues who have had past success as they pursue large external competitive grants. As an individual submitting external competitive grants, starting with relatively small budget requests will help you develop the expertise and track record to compete later for larger grants. Mentors and the department head/chair can be very helpful in explaining the balance of funding sources expected for tenure, and what type of funding should be pursued at what point in your career. For example, competitive grants from federal agencies are typically a high priority for faculty with a primary research appointment. Also, junior faculty usually cannot afford to take on projects or testing agreements for which there is little to no chance for publishing. A final point about grants is quality. With the highly competitive nature of grant funding, it is critical that any faculty member puts his or her "best foot forward" with any grant proposal submission, and work hard to ensure its quality. This requires taking advantage of grantsmanship workshops, working with mentors, and completing proposals (usually requiring many drafts) with enough time to seek internal review.

Grant funds, whether focused on teaching, research, or extension efforts, help lead to the publications expected for promotion and tenure. Mentors and the department head/chair of junior faculty can be of great help regarding recommendations on what and where to publish, and whose names are appropriately included on the publication. For example, junior faculty needs to focus primarily on peer-reviewed publications appropriate for his or her appointment and discipline, rather than writing book chapters. Also as an example, the lists of authors on publications need to show that the junior faculty member is independent from the former advisor.

Preparing the Promotion Document

In addition to obtaining the university promotion and tenure policy document and the instruction for preparation of the promotion document, the following are some recommendations to help you manage the promotion process:

- Obtain a copy of promotion documents from your department head/chair for several recently promoted associate professors at your university, ideally in your own department (especially those with similar nature of responsibilities, i.e., primary teaching, research, or extension appointments)
- Get a copy of any evaluation form used during review of promotion documents at the department and/or college/school level
- Meet with your department head/chair to discuss the promotion process and yearly performance evaluation, to understand the relationship between promotion requirements and yearly evaluations
- Attend any meetings held by the department head/chair or dean about the promotion and tenure process

- After your first year as a faculty member, create your promotion document, following the required format (i.e., to help you see where the "holes" are)
- Create a plan/goal timeline for products, etc., to help plan your career (e.g., number of papers published by year)
- Update the promotion document yearly (compare document from year to year, to note areas of progress, and areas lacking in progress)
- Keep a folder/record of activities, etc. that will be included in your promotion document; Add things to this folder throughout the year, for easy inclusion in your promotion document yearly; Even better, keep all of your materials updated in your promotion document, all the time (Diamantes, 2002)
- Have the promotion document you are developing reviewed by the departmental committee at least yearly; Meet with the department head/chair after each review, to get feedback either formally or informally (Diamantes, 2002)

Doing the "Right" Things Well

While attention to the promotion and tenure document and process is important, focusing only on this is unhealthy and unwise. Do not think of everything you do in terms of whether it is important for promotion and tenure. Do not be obsessed with getting promoted and obtaining tenure. These will come if you are doing the right things and doing them well. This means that you need to prioritize and focus on the "right" things. Advice from your department head/chair and mentors, along with knowledge of expectations for the promotion document, will help you determine what those "right" things are. A periodic self-evaluation of how you spend your time and matching these to the "right" things will help direct your prioritization and focus. Doing those "right" things well necessitates seeking help to make improvements in your skills linked to the "right" things. Examples include grant writing workshops and graduate student mentoring programs to help your research program, teaching technique, and peer evaluation workshops to help your teaching skills, and media relations workshops to help your extension program.

Being a Good Team Player

While some junior faculty members work toward promotion and tenure, they focus too much on doing the "right" things for themselves, and they neglect being a team player and "good citizen" in the department. Every faculty member in the department needs to show support to the department, the department head/chair, other faculty members, staff, and students. One example is participation in important department events/activities. Learn what events are expected/required for attendance, and make these of highest priority. Being a team player inside and outside the department is important for continued success. You want to be someone with whom other faculty members like to work. You want to become someone that people can count on for a quality and timely job on things—whether it is to write your part of a joint grant proposal, review your student's manuscript draft, serve on a graduate student's advisory committee, give a guest lecture, etc. It is important to accept responsibility as appropriate, then follow through in a timely manner and do the job well. This recommendation about being a team player is consistent with the argument by Mawdsley (1999) to junior faculty members that collegiality should be considered a major factor in tenure and promotion decisions, since "successful governance of the academic business of the university depends on cooperation." While not being liked by colleagues may not hold up in courts as a reason for tenure denial, personality and interpersonal relations can impact evaluations (Ross, 1987).

Summary

Junior faculty seeking promotion and tenure are best served by understanding the expectations for scholarly activities, learning the tenure and promotion policy and document requirements on their campus, and taking full advantage of all resources available on their campus. In addition, junior faculty need to get a strong start to build their programs, set short-term and long-term goals, begin preparing their promotion document early and get continual feedback. Final recommendations are to focus on the right things and do them well, and to be collegial and a good team player. Understanding the promotion process and focusing on the right things to build your case for promotion and tenure will greatly reduce the fear and anxiety commonly associated with this step in a successful academic career.

Acknowledgments Appreciation is extended to the many heads/chairs of food science departments across the United States, who reviewed this chapter and offered valuable suggestions to ensure wide applicability.

References

Abbott, D., and Sanders, G.F. 1991. On the road to tenure. Family Relations 40:106–109.

Boice, R. 2000. *Advice for New Faculty Members*. Needham Height, MA: Allyn and Bacon.

Boyer, E.L. 1990. *Scholarship Reconsidered: Priorities for the Professoriate*. Princeton, NJ: Carnegie Foundation for the Advancement of Teaching.

Diamantes, T. 2002. Promotion and tenure decisions using the Boyer model. Education, 00131172, Winter 2002, 123(2). Database: Academic Search Premier pp. 322–325, 333.

Diamond, R.M., 1995. *Preparing for Promotion and Tenure Review*. Bolton, MA: Anker.

Diamond, R.M. 2002. The mission-driven faculty reward system. Ch. 17 in *Field Guide to Academic Leadership*. R.M. Diamond, Ed., and B. Adam, Asst. Ed. San Francisco, CA: Jossey-Bass, John Wiley & Sons, Inc. p. 271–291.

Gelmon, S., and Agre-Kippenhan, S. 2002. Promotion, tenure, and the engaged scholar: Keeping the scholarship of engagement in the review process. AAHE Bulletin, 54(5) pp. 7–11.

Mawdsley, R.D. 1999. Collegiality as a factor in tenure decisions. Journal Personnel Evaluation in Education 13(2):167–177.

Miller, R.I. 1987. Evaluating teaching: The role of student ratings, Ch. 3 in *Evaluating Faculty for Promotion and Tenure*, p. 31–55. Evaluating scholarship and service, Ch. 4 in *Evaluating Faculty for Promotion and Tenure*, p. 56–70. San Francisco, CA: Jossey-Bass, John Wiley & Sons, Inc.

Murphy, M. 1985. A descriptive study of faculty tenure in baccalaureate and graduate programs in nursing. Journal of Professional Nursing 1:14–22.

National Education Association. 1994. *Entering the Profession: Advice for the Untenured*. Washington, DC: National Education Association.

Olswang, S.G., Cameron, C.A., and Kamai, E. 2001. The new tenure. Paper presented at American Association for Higher Education's Conference on Faculty Roles and Rewards, Tampa, FL, February, 2001.

Ross, A. 1987. Tenure or the great chain of being Academic life and the wheel of fortune. Change 19(4): 54–55.